Membrane Biofilm Reactor Process for Nitrate and Perchlorate Removal

The mission of the Awwa Research Foundation (AwwaRF) is to advance the science of water to improve the quality of life. Funded primarily through annual subscription payments from over 1,000 utilities, consulting firms, and manufacturers in North America and abroad, AwwaRF sponsors research on all aspects of drinking water, including supply and resources, treatment, monitoring and analysis, distribution, management, and health effects.

From its headquarters in Denver, Colorado, the AwwaRF staff directs and supports the efforts of over 800 volunteers, who are the heart of the research program. These volunteers, serving on various boards and committees, use their expertise to select and monitor research studies to benefit the entire drinking water community.

Research findings are disseminated through a number of technology transfer activities, including research reports, conferences, videotape summaries, and periodicals.

Membrane Biofilm Reactor Process for Nitrate and Perchlorate Removal

Prepared by:
Samer Adham, **Thomas Gillogly**, and **Geno Lehman**
MWH
300 North Lake, Suite 1200, Pasadena, CA 91101

and

Bruce Rittmann and **Robert Nerenberg**
Northwestern University
2145 Sheridan Road, Evanston, IL 60208

Jointly sponsored by:
Awwa Research Foundation
6666 West Quincy Avenue, Denver, CO 80235-3098

U.S. Environmental Protection Agency
Washington, D.C.

and

East Valley Water District
San Bernardino, CA

Published by:

Awwa Research Foundation
Advancing the Science of Water®

IWA Publishing

DISCLAIMER

This study was funded by the Awwa Research Foundation (AwwaRF). AwwaRF assumes no responsibility for the content of the research study reported in this publication or for the opinions or statements of fact expressed in the report. The mention of trade names for commercial products does not represent or imply the approval or endorsement of AwwaRF. This report is presented solely for informational purposes.

Published by IWA Publishing, Alliance House, 12 Caxton Street, London SW1H 0QS, UK

Telephone: +44 (0) 20 7654 5500; Fax: +44 (0) 20 7654 5555; Email: publications@iwap.co.uk;
Web: **www.iwapublishing.com**

AwwaRF report number 91004F. Originally published by AwwaRF for its subscribers in 2003.
IWA Publishing version published 2004.

ISBN 1 84339 8990

CONTENTS

TABLES

FIGURES

FOREWORD

The Awwa Research Foundation is a nonprofit corporation that is dedicated to the implementation of a research effort to help utilities respond to regulatory requirements and traditional high-priority concerns of the industry. The research agenda is developed through a process of consultation with subscribers and drinking water professionals. Under the umbrella of a Strategic Research Plan, the Research Advisory Council prioritizes the suggested projects based upon current and future needs, applicability, and past work; the recommendations are forwarded to the Board of Trustees for final selection. The foundation also sponsors research projects through an unsolicited proposal process; the Collaborative Research, Research Applications, and Tailored Collaboration programs; and various joint research efforts with organizations such as the U.S. Environmental Protection Agency, the U.S. Bureau of Reclamation, and the Association of California Water Agencies.

This publication is a result of one of these sponsored studies, and it is hoped that its findings will be applied in communities throughout the world. The following report serves not only as a means of communicating the results of the water industry's centralized research program but also as a tool to enlist the further support of the nonmember utilities and individuals.

Projects are managed closely from their inception to the final report by the foundation's staff and large cadre of volunteers who willingly contribute their time and expertise. The foundation serves a planning and management function and awards contracts to other institutions such as water utilities, universities, and engineering firms. The funding for this research effort comes primarily from the Subscription Program, through which water utilities subscribe to the research program and make an annual payment proportionate to the volume of water they deliver and consultants and manufacturers subscribe based on their annual billings. The program offers a cost-effective and fair method for funding research in the public interest.

A broad spectrum of water supply issues is addressed by the foundation's research agenda: resources, treatment and operations, distribution and storage, water quality and analysis, toxicology, economics, and management. The ultimate purpose of the coordinated effort is to assist water suppliers to provide the highest possible quality of water economically and reliably. The true benefits are realized when the results are implemented at the utility level. The foundation's trustees are pleased to offer this publication as a contribution toward that end.

Edmund G. Archuleta, P.E.
Chair, Board of Trustees
Awwa Research Foundation

James F. Manwaring, P.E.
Executive Director
Awwa Research Foundation

ACKNOWLEDGMENTS

This project team is indebted to the East Valley Water District for initiating the process to secure federal funding for these projects and for providing general project oversight.

The authors would like to acknowledge the financial and administrative support that has been provided by the U.S. Federal Government through the Environmental Protection Agency. It should be noted the information presented herein is based on the views of the project team and not of Awwa Research Foundation or the U.S. Federal Government.

The authors wish to thank Traci Case, project manager, of AwwaRF and the following members of the Project Advisory Committee; Kevin Mayer, U.S. Environmental Protection Agency, San Francisco, CA; John Carman, Metropolitan Water District of Salt Lake City & Sandy, Utah; and Robin Collins, The University of New Hampshire, Durham, NH; for their guidance and suggestions.

The authors of this report are indebted to the La Puente Valley County Water District for its involvement and hosting the pilot testing.

The authors would also like to thank Mitsubishi Rayon Corporation for their continuing support in this project by providing some of the membrane modules for pilot testing.

Finally he authors would like to acknowledge the effort Dr. Yasunori Kawagoshi, his support from the Osaka Metropolitan Government, and the work of Darren Giles, Brian Gallagher, Eric Bruce and Kristie Witter from the MWH Applied Research Department. The insight and comments from Amy Zander, Clarkson University, Potsdam, NY; and Rhodes Trussell, MWH, Pasadena, CA, were also appreciated. In addition to the above, quality assurance was also provided by Joan Oppenheimer, MWH, Pasadena, CA.

EXECUTIVE SUMMARY

Since the detection of perchlorate in some California groundwaters in early 1997, several studies have evaluated multiple treatment technologies for its removal. Several treatment technologies have proven to be technically feasible at the drinking-water treatment scale: biological reduction, ion-exchange, reverse osmosis (RO) membranes, and granular activated carbon (GAC). Although ion exchange and RO are proven technologies for water treatment, they only concentrate perchlorate into waste streams that require further disposal and/or treatment. Commercially available GAC has an exceptionally limited perchlorate treatment capacity. On the other hand, biological reduction reduces perchlorate to the innocuous chloride ion (Cl^-) without the production of any residuals that require special handling, which is a major advantage.

Numerous approaches and process configurations have been utilized to develop a biologically reductive system. The system presented herein centers around a novel membrane biofilm reactor (MBfR). This reactor contains hollow-fiber membranes potted at both ends of the cylindrical module. Hydrogen is fed to one end of the reactor, filling the inside of the fibers and passively diffuses through the membranes to serve as an electron donor for the biofilm that grows on the outside of the hollow-fibers. Hydrogen is an ideal electron donor, as it is non-toxic, inexpensive, and sparsely soluble. The biofilm within the reactor was developed from the indigenous bacteria present in the groundwater and was not artificially inoculated or amended.

In this research, initial feasibility investigations with the MBfR were performed at bench-scale to demonstrate its ability to reduce perchlorate. A major difference between the biological reduction of perchlorate and the more familiar biological denitrification processes is that perchlorate is typically present at concentrations orders of magnitude lower than nitrate. Consequently, it may be reduced as a secondary substrate, with another electron acceptor (nitrate or oxygen) serving as the primary substrate. For this to happen, perchlorate-reducing bacteria (PCRB) would likely be a subset of the denitrifying and oxygen-reducing microbial populations within the reactor.

During the initial investigations, the reduction of perchlorate to 4 μg/L from 60 to 1,000 μg/L was observed. No specialized inoculum for the MBfR was required, although a two-week adaptation period was required before removals of 99 percent were observed. The lag was presumably the time required to enrich for the perchlorate-reducing population. It was also discovered that the perchlorate-reducing ability of the MBfR was more sensitive to environmental conditions than its nitrate-reducing ability.

Once it was established that the MBfR could be used to degrade perchlorate, kinetic studies were performed to better understand the fundamentals of perchlorate reduction under autotrophic hydrogen-oxidizing conditions. These tests were performed with a pure culture of *Dechloromonas* sp. PC1, an autotrophic hydrogen-oxidizing bacterium isolated from the MBfR. PC1 was characterized in terms of its phylogeny, physiology, and perchlorate-reduction kinetics. Chlorate, the first reduction intermediate from perchlorate, inhibits perchlorate competitively, a novel finding for a PCRB. When a high perchlorate concentration was reduced, enough chlorate accumulated to slow the perchlorate reduction rates. However, at low perchlorate concentrations, chlorate did not accumulate enough to have an effect on perchlorate reduction rates. Nitrate also had an inhibitory effect on perchlorate reduction, but appeared to be non-competitive and constant. PC1's half-maximum rate concentration (K) for perchlorate was 0.15 mg/L, around two orders of magnitude lower than found for other PCRB. Based on the kinetic parameters for perchlorate reduction, the estimated growth threshold for perchlorate (S_{min}), for autotrophic, hydrogen-oxidizing conditions, was approximately 5 μg/L. This means that perchlorate can not serve as a primary acceptor in a completely mixed, steady-state reactor with a 4-μg/L effluent perchlorate concentration, the California Department of Health Services (CaDHS) Action Level. This finding underscored the need to have nitrate or oxygen present as a primary acceptor.

Further investigations into the microbial ecology of the mixed cultures in the MBfRs revealed that PCRB were found to be present in a denitrifying system that had not been previously exposed to perchlorate. However, a dominant PCRB species increased from 14 to 21 percent of total bacteria when 100-μg/L perchlorate was added to the influent. Increasing perchlorate reduction led to further increases in the dominant PCRB and the perchlorate-removal capacity of MBfRs. Another important finding was that oxygen alone can serve as a primary acceptor for perchlorate reduction, and that the oxygen reduction appeared to be more favorable for perchlorate reduction than was nitrate reduction.

Scaling up the MBfRs by utilizing full-scale hollow-fiber membrane modules as the core of the pilot plant was the first step in understanding the scale-up, long-term operational, and maintenance issues associated with the operation of an MBfR. Also included in the pilot plant were an aeration basin to re-oxygenate the water after the MBfR and a dual-media filter to capture any sloughed biomass from the reactors. A total of four different MBfR reactor designs were evaluated as a part of this study. Through the testing of these reactors, it was determined that the two most important factors affecting the long-term efficacy of the system were: 1) the mass-transport of perchlorate to the biofilm surface; and 2) the effective control and removal of excess biomass.

Changes to perchlorate mass-transport to the biofilm surface were realized by altering the water influent and effluent configurations, recycling water through the reactor, altering membrane fiber packing density, baffling, and removing excess biomass (to prevent channeling).

The effective control and removal of excess biomass (i.e., cleaning the MBfR) was achieved through increased linear velocity through the reactor (recycle flow), increasing fiber mobility (only pot fiber on one end of the MBfR), regular air scour (daily), and periodic mild acid cleanings (monthly). These cleaning procedures were able to control the build-up of excess biomass that can lead to clumping of the fibers thereby, reducing the effective biofilm area, exacerbating short-circuiting, reducing mass transport of perchlorate to the reductive organisms in the biofilm, and increasing the distance hydrogen must diffuse to reach the outer portion of the biofilm. Regular cleaning was also effective in minimizing sulfate reduction by impeding the proliferation of slow-growing sulfate-reducing bacteria. Additionally, the cleanings were shown to be an effective mechanism to reduce fiber breakage as a result of the calcification of the membrane fibers, or more specifically, the biofilm surrounding the membrane fibers.

Based on the knowledge acquired throughout the testing program, the anticipated cost of a full-scale MBfR based treatment system was compared to an existing perchlorate removal system. A wide range in the unit water production treatment costs associated with the scenarios evaluated was discovered. Perchlorate treatment (amortized capital, operation & maintenance) at the existing La Puente Valley County Water District 2500 gpm facility in southern California costs \$1.09/1000 gal (\$356/acre-ft). This system only removes perchlorate and discharges it into a dedicated brine line. Once perchlorate-laden brine discharge is no longer allowed, the additional costs associated with destruction of the perchlorate in the brine is anticipated to increase the costs of treatment to approximately \$1.32/1000 gal (\$430/acre-ft). While this represents a 21-percent increase in the cost of treatment, it is lower than the cost calculated for the initial MBfR designs used directly in the pilot study. However, projected modifications based on the results of the bench- and pilot-scale studies, along with good engineering, make the cost of an MBfR system that destroys perchlorate cost competitive at around $1.10/1000 gal ($360/acre-ft).

In addition to a technological and economic evaluation of the technology, multiple state agencies responsible for potential regulatory approval of the MBfR treatment process were contacted to obtain their comments about the MBfR process for the production of potable quality water. A variety of comments were received; however, the overall response summarized by the California Department of Health Services was that the, "membrane bio[film]reactor does appear to be a promising technology for perchlorate reduction. The Department looks forward to future communications from [the project team] regarding this technology."

CHAPTER 1
INTRODUCTION

Since the discovery of perchlorate (ClO_4^-) contamination in a number of California groundwaters in 1997, it has been detected in many other locations across the country. The United States Environmental Protection Agency (USEPA, 2003) estimates that groundwaters in at least 44 states have the potential to be contaminated with perchlorate (Logan, 2001), and has confirmed perchlorate releases in 25 states (Figure 1.1), with California having the highest number of confirmed releases (USEPA, 2003; Mayer, 2003). It appears to be linked to the historical manufacturing, usage, or processing of ammonium perchlorate (NH_4ClO_4), a solid rocket fuel. In March 1998, the USEPA formally added perchlorate to the drinking water contaminant candidate list (CCL) (Perciasepe, 1998). Its monitoring in drinking water supplies was mandated in 1999 under the Unregulated Contaminants Monitoring Rule (UCMR) (Browner, 1999).

The primary concerns over perchlorate toxicity are based on its interference of iodide uptake by the thyroid gland, as well as the related possible carcinogenic, developmental, reproductive, and immunotoxic effects resulting from this interference. In adults, the thyroid helps to regulate metabolism. In addition to metabolism, the thyroid plays a major role in proper development of children. Impairment of thyroid function in expectant mothers may impact the fetus and newborn and result in effects including changes in behavior, delayed development and decreased learning capabilities. Changes in thyroid hormone levels may also result in thyroid gland tumors (USEPA, 2003).

The California Department of Health Services (CaDHS) first established an action level in 1997 when, in cooperation with the Office of Environmental Health Hazard Assessment (OEHHA), it reviewed USEPA's 1992 and 1995 evaluations of perchlorate. USEPA, as part of its Superfund activities, had developed a "provisional" reference dose (RfD) for perchlorate, based on the chemical's effects on the thyroid gland. CaDHS established an 18-µg/L action level, which corresponded to the upper value of the 4- to 18-µg/L range that resulted from USEPA's provisional RfD.

EPA released for public review and comment its revised draft toxicity assessment on perchlorate entitled, "Perchlorate Environmental Contamination: Toxicological Review and Risk Characterization," in January 2002 which specified a revised draft RfD of 1 µg/L. Based on this, CaDHS concluded that its perchlorate action level needed to be revised downward. Accordingly, on January 18, 2002, CaDHS reduced the perchlorate action level to 4 µg/L, the lower value of the 4- to 18-µg/L range that resulted from the earlier provisional RfD. In December 2002, the California Office of Environmental Health Hazard Assessment (OEHHA) released a revised draft perchlorate Public Health Goal (PHG) proposing a concentration of 2 to 6 µg/L. OEHHA's PHG, when final, contributes to DHS' development of an MCL for perchlorate. Currently, CaDHS continues to utilize the 4-µg/L action level.

PRIMARY PROJECT OBJECTIVES

As a collaborative effort between MWH and Northwestern University, the application of a bioreactor system to low-concentration perchlorate-contaminated water was investigated. The

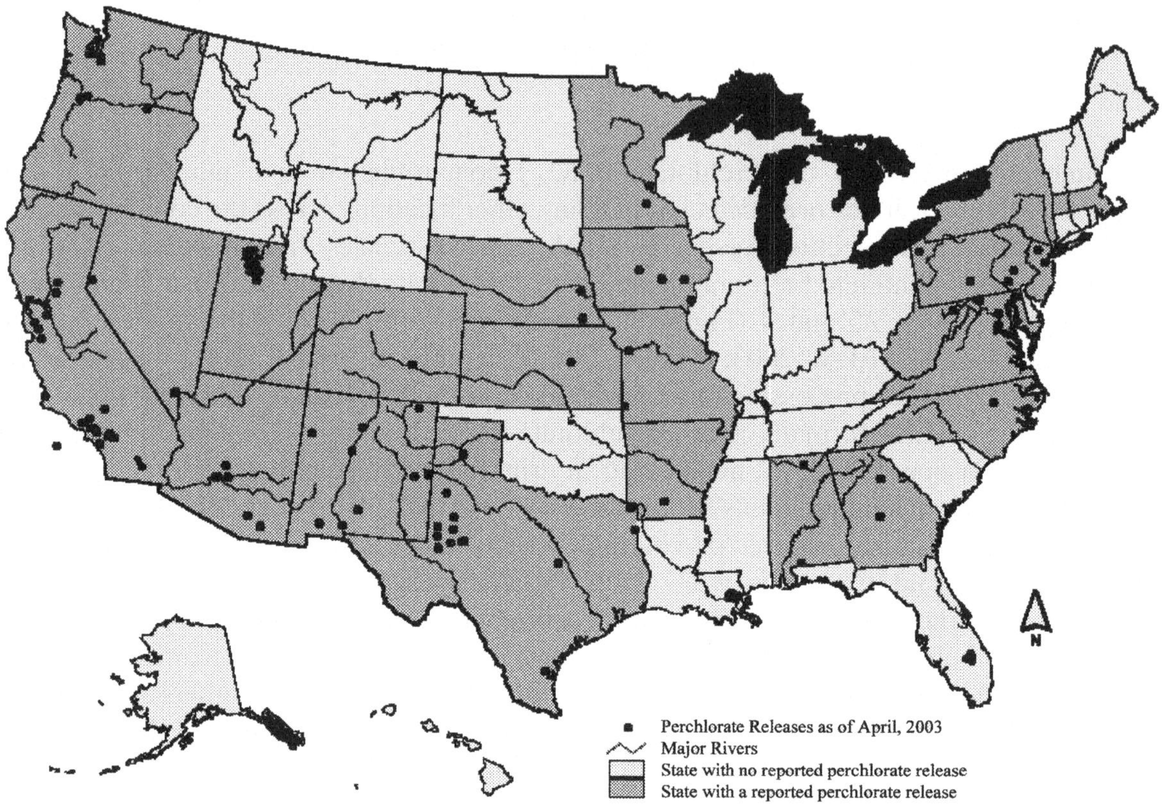

Figure 1.1 Perchlorate releases as of April 2003 (USEPA, 2003)

system was built around a novel hydrogen-fed hollow-fiber membrane biofilm reactor (MBfR). The primary objectives of the project focused on:

- *MBfR Feasibility.* Through initial bench-scale investigations, establish that the MBfR is a feasible method of treating perchlorate-contaminated water.
- *Kinetics of Perchlorate Reduction.* Determine kinetic parameters for a pure culture under autotrophic, hydrogen-oxidizing conditions to better understand the fundamentals of perchlorate reduction.
- *Perchlorate Reducing Microbial Ecology.* Characterize the organisms present in the MBfRs and attempt to understand how and why perchlorate-reducing bacteria proliferate in these systems.
- *Process Scale-Up and Performance.* Develop a pilot-scale system based on the project team's understanding of the bench-scale reactor. The pilot-scale system will be used to evaluate reactor design, operation, and process performance.

SECONDARY PROJECT OBJECTIVES

Secondary objectives that were addressed during this project included:

- Evaluate the unit operations downstream the MBfR that would be necessary for the production of potable quality water
- Ascertain issues that might need to be addressed before regulatory approval of the process for potable water production could be granted
- Investigate the effect of different membrane materials
- Develop preliminary design criteria for a full-scale system and evaluate costs;
- Identify full-scale operational and maintenance issues
- Estimate full-scale MBfR capital, operation, and maintenance costs
- Compare MBfR costs to an existing perchlorate-removal system

CHAPTER 2
FUNDAMENTALS OF BIOLOGICAL PERCHLORATE REDUCTION

This chapter provides a review of the fundamentals of cell metabolism and biodegradation, followed by a review of denitrification and dissimilatory perchlorate-reduction processes. Finally, concurrent denitrification and perchlorate reduction is discussed.

CELL METABOLISM AND BIODEGRADATION

Cell metabolism refers to the set of biochemical anabolic and catabolic reactions that occur within a cell. Anabolic reactions are those by which basic cell constituents (e.g., proteins, lipids, and sugars) are built from nutrients (e.g., carbon, nitrogen, phosphorus, sulfur, metals). Many anabolic reactions are thermodynamically unfavorable; they must be fueled by other favorable reactions. Catabolic reactions are energetically favorable redox reactions used by bacteria to obtain and store energy. Reductions of nitrate or perchlorate normally fall into the catabolic category.

Figure 2.1 provides a schematic representation of electron and energy flows in bacterial catabolism. The electron donor and acceptor substrates are key in "fueling" the overall metabolic process. If one or the other substrate is missing, metabolism cannot occur (Madigan, Martinko, and Parker, 1997; Rittmann and McCarty, 2001). As shown in the figure, an electron donor substrate (D) is oxidized, releasing two electrons. The electrons are transferred to an oxidized internal carrier (IC), reducing it to ICH_2. The internal carriers convey electrons to other parts of the cell. A fraction of the electrons, f_e, is used for reduction of electron acceptor substrates (A), transforming the low-energy form of the internal energy carrier, adenosine diphosphate (ADP), to the high energy form, adenosine triphosphate (ATP). The remainder is used for biomass synthesis. ATP is then used to fuel anabolic reactions or for cell maintenance. A specific enzyme catalyzes each of these electron transfers.

Environmental contaminants may play several roles in cell metabolism (Rittmann, 1992; Rittmann and Sáez, 1993). The most basic role is as an electron donor or acceptor, where they provide the bacteria energy for growth. Rittmann and Sáez (1993) provide a detailed discussion on the roles substrates may play in cell metabolism, summarized below.

Primary Substrates

Primary substrates are those that provide sufficient flows of electrons and energy for sustained cell metabolism, i.e., they provide the bulk of the energy for cell maintenance and growth requirements (Rittmann and Sáez, 1993). The concept of primary substrate applies to electron-donor substrates and electron-acceptor substrates. The primary electron-donor substrates may be a reduced organic compound, such as acetate or methanol (for heterotrophic bacteria), or inorganic compounds, such as hydrogen or reduced sulfur (for autotrophic bacteria). In some cases, the primary substrate may also provide nutrients, such as carbon and nitrogen. Common primary electron acceptors are oxygen, nitrate, nitrite, sulfate, or carbon dioxide, although many other organic and inorganic compounds may also serve as electron acceptors (Madigan, Martinko, and Parker, 1997).

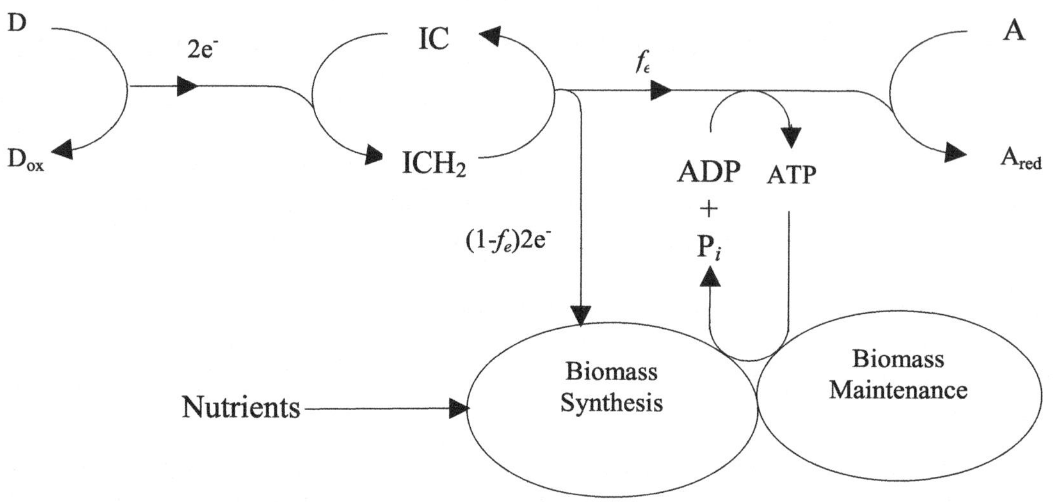

Figure 2.1 Electron and energy flows in a bacterial cell during metabolism (adapted from Rittmann and Sáez, 1993)

Secondary Substrates

Secondary substrates are those oxidized or reduced with little or no contribution to electron or energy flow. Bacteria can degrade these compounds on a sustained basis only if a primary substrate supports the biomass. Secondary substrates fall into two categories: low-concentration secondary substrates and classical co-metabolites.

Low-concentration secondary substrates are those that contribute electrons and energy flows to cell metabolism, but their concentrations are too low to be the sole primary substrate (Rittmann and Sáez, 1993). The S_{min} concept, developed for biofilms or suspended biomass (Rittmann and McCarty, 2001), defines the lowest concentration of a substrate that allows it to behave as a primary substrate. If the substrate concentration is less than S_{min}, biomass will disappear over time.

Unlike low-concentration secondary substrates, classical cometabolites are compounds whose biotransformation does not generate energy flow, even when present at high concentrations (Alexander, 1981; Dalton and Stirling, 1982; Rittmann and Sáez, 1993). Co-metabolism typically occurs when the enzymes for a primary substrate lack specificity and incidentally transform the co-metabolites; oxidized compounds may also act as a classic co-metabolite.

Modulators of Biodegradation Kinetics

Some substrates may affect the biodegradation kinetics of other substrates, known as modulators. The three common modulators are co-substrates, competitive inhibitors, and inducers (Rittmann and Sáez, 1993).

Co-substrates are compounds that react with the main substrate or with an intermediate of the main substrate at the enzyme reactive site. Most co-substrates can also play the role of primary electron donor or acceptor substrates, but lack the ability to produce respiratory electron flow.

Inhibitors slow enzyme-catalyzed reactions by reacting with the enzyme. Depending on how the inhibitor reacts with the enzyme, the inhibition may be competitive, non-competitive, or a combination of the two. A competitive inhibitor is usually an analog of the target substrate and competes with the substrate for enzyme site. A non-competitive inhibitor complexes with the enzyme at a non-reactive location, so that the enzyme's efficiency with the target substrate is reduced.

Inducers signal the cells to form of enzymes that carry out the transformations of some target compounds. Even if the bacteria possess the genetic material to express enzymes that degrade the target compound, an inducer must be present in order to initiate transcription of the DNA when the enzyme is inducible. When enzymes are present at all times, they are called constitutive.

DISSIMULATORY PERCHLORATE REDUCTION

Certain types of bacteria can use perchlorate as an electron acceptor for anaerobic growth. Such bacteria appear to be ubiquitous in the environment, even though perchlorate is almost exclusively released from human activities and has been produced only for about 100 years (Urbansky, 1998). Since all identified perchlorate-reducing bacteria also reduce chlorate, they are sometimes referred to as "(per)chlorate"-reducing bacteria (PCRB) to reflect their ability to use both these acceptors.

The existence of PCRB has been known for over 50 years. One of the first studies documenting bacterial growth on chlorate was by Bryan and Rohlich (1954). Based on the observation that chlorate could be bacterially reduced to chloride using organic matter as an electron donor, they proposed measuring chloride production from chlorate as a surrogate test for biochemical oxygen demand (BOD). In the 1970s, Korenkov et al. (1976) patented a process to treat perchlorate- and chlorate-containing wastewaters by mixing them with municipal wastewaters. In the early 1990s, interest in chlorate-reducing bacteria was stimulated by the need to treat perchlorate- or chlorate-containing industrial wastes, such as kraft mill effluent (Malmqvist, Welander, and Gunnarsson, 1991; Malmqvist and Welander, 1992), match factory wastewater (Stepanyuk et al., 1992), and rocket manufacture wastewaters (Attaway and Smith, 1993). Today, using perchlorate-reducing bacteria as a means to remediate groundwater and treat drinking water is of intense interest (Herman and Frankenberger Jr., 1998; Logan and Kim, 1998; Urbansky, 1998; Coates et al., 1999; Nerenberg, Rittmann, and Najm, 2002). PCRB appear to have many similarities to denitrifying bacteria. These similarities are highlighted as appropriate in the following sections.

Ubiquity of (Per)chlorate-Reducing Bacteria

PCRB have been found in environments containing (per)chlorate, including paper mill wastewater (Bruce, Achenbach, and Coates, 1999) and match factory wastewater (Stepanyuk et al., 1992). However, they also have been found in numerous environments that probably never contained a significant amount or any (per)chlorate. Such environments include municipal activated sludge, anaerobic digester sludge, river and ditch water, soils and sediments, "pristine" environments, and various sites with industrial contamination (Bryan and Rohlich, 1954; Malmqvist, Welander, and Gunnarsson, 1991; Malmqvist et al., 1994; van Ginkel, Plugge, and Stroo, 1995; Rikken, Kroon, and van Ginkel, 1996; Herman and Frankenberger, 1999; Attaway and Smith, 1993; Coates et al., 1999). Table 2.1 summarizes environments in which PCRB have been found.

Table 2.1
(Per)chlorate reducing bacteria in the environment

Environment	Reference
Match factory wastewater	Stepanyuk et al. (1992)
Paper mill wastewater	Bruce, Achenbach, and Coates (1999)
Municipal activated sludge	Bryan and Rohlich (1954); Malmqvist, Welander, and Gunnarsson (1991); Malmqvist et al. (1994); van Ginkel, Plugge, and Stroo (1995); Rikken, Kroon, and van Ginkel (1996); Herman and Frankenberger (1999)
Anaerobic digester sludge	Attaway and Smith (1993); van Ginkel, Plugge, and Stroo (1995)
Granular sludge from a sugar mill upflow anaerobic sludge blanket	van Ginkel, Plugge, and Stroo (1995)
Swine lagoon	Coates et al. (1999)
River and ditch water, soils, and sediments	van Ginkel, Plugge, and Stroo (1995); Coates et al. (1999)
Gold mine drainage sediment	Coates et al. (1999)
Petroleum-contaminated sediment	Coates et al. (1999)

Diversity of (Per)chlorate-Reducing Bacteria

Although relatively few PCRB have been isolated and characterized at this time, the available information suggests that, like denitrifying bacteria, they are phylogenetically and phenotypically diverse. Phylogenetically, PCRB are widespread among the *Proteobacteria* (Malmqvist et al., 1994; Rikken, Kroon, and van Ginkel, 1996; Wallace et al., 1996; Bruce, Achenbach, and Coates, 1999; Coates et al., 1999; Herman and Frankenberger, 1999; Zhang, Bruns, and Logan, 2002). Known isolates span four of the five *Proteobacteria* subgroups, including the alpha, beta, gamma, and epsilon subgroups. Most of these known isolates, however, are members of the beta subgroup of *Proteobacteria* and of the genus *Dechloromonas* or *Dechlorosoma* (Bruce, Achenbach, and Coates, 1999; Coates et al., 1999; Achenbach et al., 2001; Logan et al., 2001; Zhang, Bruns, and Logan, 2002).

Phenotypically, most PCRB are Gram-negative, non-spore-forming, non-fermenting, cytochrome-c positive, facultative anaerobes (Bruce, Achenbach, and Coates, 1999; Coates et al., 1999; Logan et al., 2001).

Table 2.2 lists some of the known (per)chlorate-reducing isolates, their phylogeny, and their electron donor and acceptors. Strain PK, from the gamma-*Proteobacteria*, is not shown in the table. The morphologies are mainly motile rods (GR-1, *I. Dechloratans*, HAP-1, perc1ace), although strain WD is a spirillum, and *I. Dechloratans* can occur in pairs or short filaments. In most cases, organic acids and alcohols were used to support perchlorate reduction (Bruce, Achenbach, and Coates, 1999; Coates et al., 1999; Herman and Frankenberger, 1999; Logan et al., 2001). An exception is HAP-1, which appears to require protein-based amendments such as yeast extract, acid-hydrolyzed casein, or peptones for perchlorate reduction.

Table 2.2
Metabolic diversity of (per)chlorate-reducing bacteria

Strain	Phylogeny	Acceptors*	Donors	Reference
Vibrio dechloratans Cuznesove		ClO_4^-, ClO_3^-, NO_3^-, O_2	Heterotrophic	Korenkov et al. (1976)
Acinetobacter thermotoleranticus		ClO_3^-, $SO_4^=$, O_2	Heterotrophic	Stepanyuk et al. (1992)
Ideonella dechloratans	β-*Proteobacteria*	ClO_3^-, NO_3^-, O_2 [NO_2^-, $SO_4^=$]	Heterotrophic	Malmqvist, Welander, and Gunnarsson (1991), Malmqvist et al. (1994)
Wolinella succinogenes HAP-1	ε-*Proteobacteria*	ClO_4^-, ClO_3^-, NO_3^-, O_2 [NO_2^-, $SO_4^=$, ClO_2^-]	Heterotrophic, mixotrophic with H_2	Wallace et al. (1996, 1998)
GR-1	β-*Proteobacteria*	ClO_4^-, ClO_3^-, NO_3^-, O_2, Mn(IV) [$SO_4^=$, IO_3^-, BrO_3^-,ClO_2^-]	Heterotrophic	Rikken, Kroon, and van Ginkel (1996)
Dechlorimonas agitatus sp. CKG	β-*Proteobacteria*	ClO_4^-, ClO_3^-, O_2 [NO_3^-, NO_2^-, SO_3^-, $SO_4^=$, ClO_2^-, Mn(IV)]	Heterotrophic, mixotrophic with Fe(II), H_2S	Bruce, Achenbach, and Coates (1999)
Isolates NM and CL	β-*Proteobacteria*	ClO_3^-, O_2 [NO_3^-, $SO_4^=$, SeO_3^-, Mn(IV), Fe(III) and others]	Heterotrophic	Coates et al. (1999, 2000)
Isolate MissR, Iso1, Iso2, SDGM	β-*Proteobacteria*	ClO_3^-, NO_3^-, O_2, [$SO_4^=$, SeO_3^-, Mn(IV), Fe(III) and others]	Heterotrophic	Coates et al. (1999, 2000)
Isolate PS	β-*Proteobacteria*	ClO_3^-, NO_3^-, O_2, [$SO_4^=$, SeO_3^-, Mn(IV), Fe(III) and others]	Heterotrophic, Fe(II)	Coates et al. (1999, 2000)
Isolate WD	α-*Proteobacteria*	ClO_3^-, NO_3^-, O_2, [$SO_4^=$, SeO_3^-, Mn(IV), Fe(III) and others]	Heterotrophic, Fe(II)	Coates et al. (1999, 2000)
Perc1ace	β-*Proteobacteria*	ClO_4^-, NO_3^-, O_2	Heterotrophic	Herman and Frankenberger (1999)
Autotrophic consortium			Mixotrophic with H_2	Giblin et al. (2000)
JM	β-*Proteobacteria*	ClO_4^-, ClO_3^-, NO_3^-, O_2, [$SO_4^=$]	Mixotrophic with H_2	Miller and Logan (2000)
PDX, KJ	β-*Proteobacteria*	ClO_4^-, ClO_3^-, NO_3^-, O_2	Lactate, acetate	Logan et al. (2001)
HZ	β-*Proteobacteria*	ClO_4^-, ClO_3^-, NO_3^-, O_2,	Autotrophic with hydrogen	Zhang, Bruns, and Logan (2002)
PC1	β-*Proteobacteria*	ClO_4^-, ClO_3^-, NO_3^-, NO_2^-, O_2, [SO_4^{2-}, BrO_3^-, SeO_4^{2-}]	Hydrogen, acetate	Nerenberg (2003)

* Negatively tested acceptors and inhibitors are shown in brackets

Several important facts are evident in the Table 2.2: all of these (per)chlorate-reducing isolates can use oxygen as an electron acceptor (i.e., are facultative anaerobes or microaerophiles similar to denitrifiers); almost all can use nitrate as an electron acceptor. Third, most are heterotrophs, although some can use Fe(II), H_2S, or H_2 mixotrophically or autotrophically. One isolate, *Acinetobacter thermotoleranticus*, can use sulfate as an electron acceptor: all of these isolates tested for perchlorate also reduced chlorate. It should be noted that some species not identified here have been found to reduce chlorate, but not perchlorate (Logan et al., 2001).

Most PCRB reduce perchlorate to chloride with no significant accumulation of intermediates. Potential intermediates of perchlorate reduction include chlorate, chlorite and hypochlorite, which are toxic to bacteria (Goksøyr, 1952; Pichinoty, 1969; Anderson et al., 2000). The elimination of intermediates may be a necessary strategy for survival.

Autotrophic bacteria are those that do not require an organic carbon source for growth. Instead, they grow with bicarbonate as a carbon source. Several studies show that mixed cultures of hydrogen-oxidizing autotrophic cultures are easily obtained (Attaway and Smith, 1993; van Ginkel, Plugge, and Stroo, 1995; Giblin et al., 2000). Only one researcher has isolated a hydrogen-oxidizing, (per)chlorate-reducing autotroph (Zhang, Bruns, and Logan, 2002). Instead, most researchers have found mixotrophic bacteria that grow with perchlorate and hydrogen only in the presence of an additional organic carbon source (Wallace et al., 1998; Giblin et al., 2000; Miller and Logan, 2000). As most ground waters are oligotrophic (i.e., have very low organic carbon content), autotrophic, hydrogen-oxidizing, PCRB would most likely be favored under such conditions. Some mixed cultures of hydrogen-oxidizing PCRB in waters devoid of carbon sources may include other bacteria that provide organic carbon to the mixotrophic bacteria. These may be methanogens, acetogens, or autotrophic bacteria that grow with other acceptors, such as nitrate. Mixotrophic, hydrogen-oxidizing, perchlorate reducing bacteria may be favored in environments with greater organic carbon content, or with abundant alternative acceptors.

Perchlorate Reduction Enzymes

Two specialized enzymes are involved in dissimulatory perchlorate reduction to chloride: (per)chlorate reductase and chlorite dismutase. The individual reactions catalyzed by each enzyme and the overall reduction reaction are shown in Table 2.3. As shown in this table, perchlorate reduction occurs in three steps. First, perchlorate is reduced to chlorate by the (per)chlorate-reductase enzyme in a two-electron transfer; second, chlorate is reduced to chlorite by the same enzyme, in another two-electron transfer; and third, chlorite is transformed to chloride and oxygen by chlorite dismutase enzyme and without any external electron transfer.

(Per)chlorate Reductase

Kengen et al. (1999) isolated from strain GR-1 an enzyme that reduces perchlorate to chlorate and chlorate to chlorite. They named the enzyme (per)chlorate reductase because it catalyzes perchlorate reduction to chlorate and chlorate reduction to chlorite. Based on activity for methyl viologen found in the whole cell fraction, Kengen et al. (1999) determined that the (per)chlorate reductase enzyme is located on the outside of the cytoplasmic membrane. Cell fractionation studies showed that the enzyme is soluble and not membrane bound. The enzyme has several similarities to nitrate reductase. It appears that the (per)chlorate reductase includes a trimer of heterodimers of 95 and 40 kDa ($\alpha_3\beta_3$), with 3 moles of molybdenum and 3 moles of

Table 2.3
Perchlorate reduction reactions

Reaction	Electrons exchanged	Change of Cl oxidation state	Free energy $\Delta G^{\circ\prime}$ (kJ/e$^-$)
Perchlorate reduction by (per)chlorate reductase $ClO_4^- + H_2 \rightarrow ClO_3^- + H_2O$	2	+7→+5	-115.9
Chlorate reduction by (per)chlorate reductase $ClO_3^- + H_2 \rightarrow ClO_2^- + H_2O$	2	+5→+3	-108.3
Chlorite reduction by chlorite dismutase $ClO_2^- \rightarrow Cl^- + O_2$	4	+3→-1	-37.1*
Oxygen reduction $O_2 + 4H_2 \rightarrow 2H_2O$	8	0→-4	-123
Theoretical total energy, without O_2 utilization† $ClO_4^- + 4H_2 \rightarrow Cl^- + 4H_2O$	8	+7→-1	-112.1
Theoretical total energy, with O_2 utilization† $ClO_4^- + 4H_2 \rightarrow Cl^- + 4H_2O$	8	+7→-1	-118

* This energy is not captured by PCRBs
† Does not include utilization of chlorite

selenium next to 30.6 moles of iron. This is similar to nitrate reductase, which has a dimer αβ with subunits of approximately 104 to 150 kDa (α) and 52 to 63 kDa (β). The α-subunit of nitrate reductase contains iron-sulfur centers and molybdenum and is catalytic. Significant differences with nitrate reductase are the periplasmic location (versus membrane bound for nitrate reductase), the presence of selenium, and the N-terminal sequence on the β-subunit.

Giblin and Frankenberger (2001) studied perchlorate- and nitrate-reductase activity for the perchlorate-reducing strain perc1ace. As in the study by Kengen et al. (1999), they found that the perchlorate-reductase activity for perchlorate-grown cells was associated with the periplasm, not with the membrane or cytoplasm. They also found that enzyme extracts from nitrate-grown cells had a small amount of perchlorate reduction activity, while perchlorate-grown cells had some nitrate reduction activity. When nitrate or perchlorate-grown resting cells were supplied with nitrate and perchlorate together, each was reduced at rates similar to those with a single acceptor. This suggests that both sets of enzymes were induced by nitrate or perchlorate.

Chlorite Dismutase

O'Connor and Coates (2002) found that the chlorite dismutase enzyme is highly conserved, even among bacteria of different phylogenetic affiliations. Two researchers studied chlorite dismutase enzymes: Van Ginkel et al. (1998) and Bruce, Achenbach, and Coates (1999). Van Ginkel et al. (1998) isolated strain GR-1 an enzyme from that catalyzed the dismutation of chlorite to chloride and oxygen. The enzyme is soluble, and, based on the periplasmic location of (per)chlorate reductase (Kengen et al., 1999), is probably located in the periplasm. Chlorite dismutase is a homotetramer composed of 32-kDa subunits, and its red color suggests it may be a heme protein. It is different from other heme enzymes, such as catalases and peroxidases, due to its molecular weight and the production of oxygen and chloride in equimolar amounts.

Bruce, Achenbach, and Coates (1999) purified and studied a chlorite dismutase enzyme from strain CKG. They found activity in the membrane and soluble fractions of the cell lysate. Although the specific activity of both fractions was similar (14.13 and 12.08 mmol chlorite converted/mg-min for the soluble and membrane fractions, respectively), the total activity was three times greater in the soluble fraction than in the membrane fraction.

Wallace et al. (1996 and 1998) discussed that HAP-1 appears to produce intermediates of perchlorate reduction that oxidize the medium. It is possible that HAP-1 lacks the chlorite dismutase enzyme, and that the oxidizing agent is actually chlorite produced from perchlorate reduction. Although chlorite was not detected, it may not accumulate because it is reduced by the yeast extract in the complex medium

Thermodynamics of Perchlorate Reduction

The thermodynamics of perchlorate reduction to chlorate and chlorate reduction to chlorite are comparable to oxygen reduction. For example, energies from perchlorate reduction to chlorate and chlorate reduction to chlorite using hydrogen as an electron donor are -115.9 kJ/electron and -108.3 kJ/electron, respectively, while the energy from oxygen reduction with hydrogen is -118.6 kJ/electron.

Although the disproportionation of chlorite to chloride cannot be coupled directly to electron transport and energy capture, the produced oxygen can be used by PCRB as an additional source of energy. Research suggests that PCRB use perchlorate and oxygen concurrently, as oxygen does not accumulate during perchlorate reduction by pure cultures of PCRB (Coates et al., 2000; Rikken, Kroon, and van Ginkel, 1996; Herman and Frankenberger, 1999). The energy obtained by perchlorate reduction to chloride with hydrogen as a donor, including oxygen utilization, is -118 kJ/electron.

Electron Transport Chain and Energy Capture

The electron transport process for PCRB is only partly understood. What is clear is that three steps are involved. The first two steps include perchlorate and then chlorate reduction by the (per)chlorate reductase enzyme, producing chlorite as a final product. Chlorite is then transformed by dismutation into chloride and oxygen. The cell may further utilize the oxygen by the regular aerobic respiratory process.

Although the reduction steps and enzymes are known, the location of these reactions in the cell and how these reductions are connected to electron transport are not clear. Kengen et al. (1999) studied the (per)chlorate reductase enzyme of strain GR-1 and found that it was located in the cell periplasm. Giblin and Frankenberger (2001) found that the (per)chlorate reductase enzyme of strain perc1ace was also located in the periplasm. This is unusual, as most respiratory enzymes are associated with the cytoplasmic membrane.

It is not known how (per)chlorate reductase is linked to electron transport, but Coates et al. (1999) found evidence that a type *c* cytochrome is involved in (per)chlorate reduction.

Table 2.4
Michaelis-Menton parameters for cell extracts

Enzyme	Substrate	V_{max} (U $Min^{-1}mg^{-1}$)	K_M (μM)	Reference
(Per)chlorate reductase	Perchlorate	3.8	27	Kengen et al. (1999)
(Per)chlorate reductase	Chlorate	13.2	<5	Kengen et al. (1999)
Chlorite dismutase	Chlorite	2,200	170	Van Ginkel et al. (1996)
Nitrate reductase	Nitrate		100 to 1300	Hochstein and Tomlinson (1988)

Another question is how electrons from reduced internal carriers are split between the (per)chlorate electron transport chain and the oxygen or transport chain. This is further complicated when nitrate is present in the medium. The mechanism by which nitrate inhibits (per)chlorate reduction may be at the level of electron flow or electron transport across the membrane. No information is available on such mechanisms at this time.

Enzyme Kinetics

Little information is available regarding the kinetic parameters for perchlorate reduction. The (per)chlorate reductase enzyme was isolated and studied by Kengen et al. (1999), and the chlorite dismutase was studied by van Ginkel et al. (1996). As shown in Table 2.4, the enzyme's activity towards chlorate is approximately three times higher than towards perchlorate. Also, kinetic data indicate a high efficiency for chlorate and perchlorate reduction, particularly as compared to nitrate reductase. Based on these parameter values, perchlorate reduction appears to be rate limiting, compared to chlorate reduction by (per)chlorate reductase and chlorite disproportionation by chlorite dismutase.

Substrate utilization rates are available for several species, although differing experimental conditions and units make them difficult to compare. *Vibrio dechloraticans Cuznove* B-1168 reduces 1.68 mg perchlorate/day-mg dry weight during anaerobic growth on acetate or ethanol (Korenkov et al., 1976). *Wolinella succinogenes* HAP-1 reduces perchlorate at a rate of 2.57 mg/day-mg dry weight in a complex medium with yeast and peptone (Wallace et al., 1998). Rikken, Kroon, and van Ginkel (1996) studied washed cell suspensions of the GR-1. The perchlorate reduction rate was 4,280 mg/g protein-hr, while the rate for chlorate was 5,670 mg/g protein-hr. In a perchlorate grown cell suspension, chloride was produced stoichiometrically from chlorite at a rate of 10-mmol min^{-1}mg $protein^{-1}$ (Rikken, Kroon, and van Ginkel, 1996). Logan et al. (2001) studied the kinetics of perchlorate and chlorate reduction for two PCRB, *Dechlorosoma* strains KJ and PDX, under non-growth conditions using acetate as an electron donor. The maximum specific growth rates were 0.055 mg/(mg protein-day) for KJ and 0.017 mg/(mg protein-day) for PDX. Converted to a dry-weight basis, assuming that protein constitutes 55% of the cell dry weight and that intracellular water constitutes 90% of the total weight (Madigan, Martinko, and Parker, 1997), the values are 24 mg/(mg DW-day) for KJ and 7.5 mg/(DW-day) for PDX. The half maximum substrate utilization rate constant, K, for perchlorate was 33 mg/L for KJ and 12 mg/L for PDX. The yields on acetate and perchlorate for KJ were 0.46 g-DW/g-acetate

Table 2.5
Monod kinetic parameters for whole cells

Isolate	Acceptor	q_{max} (mg/mgDW-day)	K (mg/L)	Reference
Vibrio dechloratans	Perchlorate	1.68		Korenkov et al. (1976)
Wolinella succcinogenes HAP-1	Perchlorate	2.57		Wallace et al. (1996, 1998)
GR-1	Perchlorate	5.65		Rikken, Kroon, and van Ginkel (1996)
GR-1	Chlorate	7.48		Rikken, Kroon, and van Ginkel (1996)
KJ	Perchlorate	24	33	Logan and LaPoint (2002)
PDX	Perchlorate	7.5	12	Logan and LaPoint (2002)

for oxygen, 0.44 g-DW/g-acetate for chlorate, and 0.5 g-DW/ g-acetate for perchlorate. The similar yields are consistent with the similar free energies per electron obtained for the three acceptors, as shown in Table 2.3. A summary of kinetic parameters is provided in Table 2.5.

Factors That Control Perchlorate Reduction

Many environmental factors may control enzyme expression and activity during perchlorate reduction. Potential factors include enzyme induction, oxygen concentration, nitrate and nitrite concentrations, (per)chlorate intermediates, temperature, and pH.

Enzyme Inhibition

(Per)chlorate-grown PCRB cease perchlorate reduction in the presence of oxygen (Wallace et al., 1996; Rikken, Kroon, and van Ginkel, 1996; Herman and Frankenberger, 1999; Chaudhuri et al. 2002), but it is not clear whether this effect results from repression or inhibition by oxygen. "Repression" is when enzyme *synthesis* is stopped, while "inhibition" is when enzyme *activity* is slowed or stopped. Oxygen inhibits perchlorate reduction in perchlorate-grown cells of *Dechlorosomas suillium*, even though chlorite dismutase was still active 5 hours after introduction of oxygen (Chaudhuri et al., 2002). Oxygen-grown cells of GR-1 were not able to reduce perchlorate or chlorate (van Ginkel et al., 1996), suggesting that oxygen is a repressor for (per)chlorate reductase in this strain. Growth with oxygen does not stop GR-1's ability to dismutate chlorite, suggesting that chlorite dismutase is a constitutive enzyme in GR-1 (van Ginkel et al., 1996). In contrast, oxygen, nitrate, or perchlorate-grown cells of strain perc1ace reduced perchlorate at similar rates (Herman and Frankenberger, 1999; Giblin and Frankenberger, 2001), suggesting that oxygen is not a repressor for perc1ace. Oxygen-grown cells of *D. suillium* (Chaudhuri et al., 2002) and *Dechloromonas agitata* strain CKB (O'Connor and Coates, 2002) did not reduce perchlorate under aerobic conditions Therefore, oxygen plays different roles in the enzyme regulation in different species of PCRB. It is likely, as with denitrification, that high levels of oxygen repress (per)chlorate reduction, while low levels only inhibit enzyme activity towards (per)chlorate.

Most PCRB can reduce nitrate as well as (per)chlorate, and some can reduce nitrite as well. Nitrate and nitrite may have an inhibitory effect on perchlorate reduction. Unfortunately, few studies have addressed this issue, and results are contradictory. One study assessed the effect of nitrate on chlorate reduction by activated sludge (van Ginkel, Plugge, and Stroo, 1995). With 835 mg/L chlorate (10 mM) and 330 mg-N/L nitrate (23 mM), nitrate was reduced, but no chlorate reduction occurred. In a separate experiment (van Ginkel, Plugge, and Stroo, 1995), nitrate added to a chlorate-reducing enrichment culture slowed chlorate reduction. For *D. suillium*, nitrate inhibited perchlorate reduction when at equimolar amounts for perchlorate-grown cells (Chaudhuri et al., 2002). These experiments suggest nitrate inhibition.

Wolinella succinogenes HAP-1 can reduce nitrate to nitrite, but cannot reduce nitrite. In a mixed culture containing this organism, approximately 9 mM of nitrate did not inhibit perchlorate reduction and was partially reduced to nitrite. However, approximately 11 mM of nitrite appeared to inhibit perchlorate reduction until it was completely reduced, and then perchlorate was reduced at a lower rate (Wallace et al., 1996). Nitrate slowed, but did not stop, perchlorate reduction by perchlorate-grown cells of *D. agitata*, a non-nitrate-reducing PCRB. Perc1ace reduced perchlorate at the same rate with and without nitrate (Giblin and Frankenberger, 2001). Thus, nitrate is not inhibitory to all PCRB, and nitrite can also inhibit (per)chlorate reduction.

Per1ace (Herman and Frankenberger, 1999) reduces nitrate and perchlorate concurrently in batch growth studies with a fixed amount of nitrate (0 or 1 mM) and varying amounts of perchlorate (1.2 mM, 0.12 mM, 0.01 mM, and 0.001 mM). At high perchlorate concentrations, nitrate reduction was slowed, but perchlorate reduction was not significantly affected. At low perchlorate concentrations, the nitrate reduction rate was increased and the perchlorate reduction rate was decreased with respect to the system with no nitrate. These studies show that perchlorate and nitrate are reduced concurrently by the same bacteria, and that inhibition effects between nitrate and perchlorate are not severe. The only significant difference for perchlorate with and without nitrate occurred at the lowest perchlorate concentration. Nitrate reduction appeared to begin earlier and end earlier than perchlorate reduction.

Nitrate and perchlorate can have mutual affects on enzyme synthesis. Perchlorate-grown GR-1 cells were able to reduce nitrate and nitrite (van Ginkel et al., 1996). Nitrate-grown cells of perc1ace had the same perchlorate reduction rates as perchlorate-grown cells, and perchlorate-grown cells had similar nitrate reduction rates as nitrate-grown cells (Herman and Frankenberger, 1999; Giblin and Frankenberger, 2001). However, nitrate-grown cells of *D. suillium* had no chlorite dismutase activity and did not reduce perchlorate (Chaudhuri et al., 2002). As with oxygen, it seems that nitrate affects induction and repression of enzymes in different ways for different bacteria.

Based on the above, it appears likely that oxygen at high levels represses (per)chlorate reductase expression, while nitrate and nitrite have varying degrees of inhibition for different strains of PCRB.

pH and Temperature

Examples of pH and temperature effects are provided in Table 2.6. For most PCRB, the pH range for (per)chlorate reduction is from 6.5 to 8.0, with an optimal between 6.8 and 7.5. The temperature range is between 25 and 40°C, typical of mesophilic bacteria.

Table 2.6
pH and temperature ranges for PCRB

Species	Optimum temp	Temp range	Optimum pH	pH range	Reference
HAP-1	40°C	20 to 45°C	7.1	6.5 to 8.0	Wallace et al. (1996)
CKG	35°C	25 and 40°C	7.5	6.5 to 8.5	Bruce, Achenbach, and Coates (1999)
Perclace	25 to 30°C	20 to 40°C	7.0 to 7.2	6.5 to 8.5	Herman and Frankenberger (1999)
Mixed culture	NA	15 to 45°C	NA	6.6 to 7.5	Attaway and Smith (1993)
Acinetobacter thermotoleranticus	36 to 37°C	4 to 47°C	6.8 to 7.2	6.0 to 7.5	Stepanyuk et al. (1992)

CONCURRENT DENITRIFICATION AND (PER)CHLORATE REDUCTION

Since PCRB can use nitrate as an electron acceptor, it is possible that they are naturally present in a mixed-culture denitrifying reactor. A trace perchlorate concentration coupled with high nitrate concentration gives only a weak selective pressure for PCRB over common denitrifiers, however. Thus, PCRB might not be present in a denitrifying system in sufficient numbers to reduce perchlorate significantly. Furthermore, it is not yet clear how much perchlorate is required to select for PCRB when the nitrate load is much higher than the perchlorate load.

In a biofilm-treatment system, the interior portions of the biofilm might provide an ideal location for PCRB, as the nitrate concentrations are much lower than the bulk liquid. Therefore, the PCRB would have a more significant advantage over non-perchlorate-reducing denitrifiers. Thus, biofilms treatment systems may be ideal for concurrent denitrification and perchlorate reduction.

CHAPTER 3
INITIAL FEASIBILITY INVESTIGATIONS

Given the ubiquity of PCRB, the favorable energy yield of perchlorate reduction, and the ability of many PCRB to respire on oxygen and nitrate, it should be "easy" to use PCRB in a biological perchlorate reduction process. Several researchers have demonstrated perchlorate reduction in reactor systems using organic donors (Wallace et al., 1998; Herman and Frankenberger, 1999; Giblin, et al., 2000; Kim and Logan, 2000a,b) or hydrogen (Giblin, Herman, and Frankenberger, 2000; Miller and Logan, 2000). However, these studies used high influent perchlorate concentrations and no nitrate, or the reactors were previously exposed to high perchlorate concentrations to enrich for PCRB. When the perchlorate concentration is low, as in most contaminated water supplies, the PCRB may not be able to compete for the electron donor with non-perchlorate-reducing bacteria, especially if other acceptors, such as oxygen or nitrate, are present at much higher concentrations.

This chapter reports on the preliminary investigations performed to assess the perchlorate reduction in a mixed-culture, denitrifying bioreactor where the reactor had no previous exposure to perchlorate, and where the nitrate concentration was much higher than that of perchlorate. Four sets of experiments were carried out. Screening experiments investigated whether biological perchlorate reduction was feasible in the drinking-water setting when hydrogen is used as the electron donor in conjunction with high nitrate loadings. Mechanism experiments identified factors influencing reactor performance. Groundwater tests evaluated treatment efficiency for an actual perchlorate-contaminated groundwater. Finally, media tests explored the effect of minimal media versus tap water on the perchlorate removal efficiency.

EXPERIMENTAL MATERIALS AND METHODS

A schematic of the MBfR used in these initial investigations is shown in Figure 3.1, and reactor characteristics are provided in Table 3.1. The reactor consisted of a membrane module connected to a recirculation loop. The system behaved as a completely mixed biofilm reactor due to the high recirculation rate. The membrane module (Porous Media, Inc., Minneapolis, MN) consisted of a bundle of hydrophobic hollow-fiber membranes inside a PVC pipe shell. The free end of each fiber was sealed so that 100 percent of the supplied hydrogen was transferred through the fiber walls to the surrounding biofilm. Each fiber was 93 cm long. The top 10 cm of each fiber was treated with a wetting agent to allow moisture condensate to exit to the bulk liquid. This portion of the membrane was much less active for hydrogen transfer than the untreated portions. Also, the bottom 10 cm of each fiber was deadened to minimize gas transfer and avoid excessive growth of biofilm that could cause the fibers to stick together. Pure hydrogen was supplied to the inside of the hollow fibers through a manifold at the base. The fibers were attached to a manifold at one end and free at the other. Water was recirculated through the looped system, causing the fibers to move independently. The water temperature in the reactor was $20 \pm 1°C$ during the experiments.

The reactor was originally inoculated in early 1998, using a pure culture of *Ralstonia eutropha*, and was then used in a denitrification study (Lee and Rittmann, 2000; Lee and Rittmann, 2001). During the denitrification study, the reactor was fed with non-sterile tap water. Therefore, the reactor presumably developed a mixed culture including hydrogen-utilizing autotrophic, denitrifying bacteria, such as *R. eutropha*, but not necessarily dominated by that species.

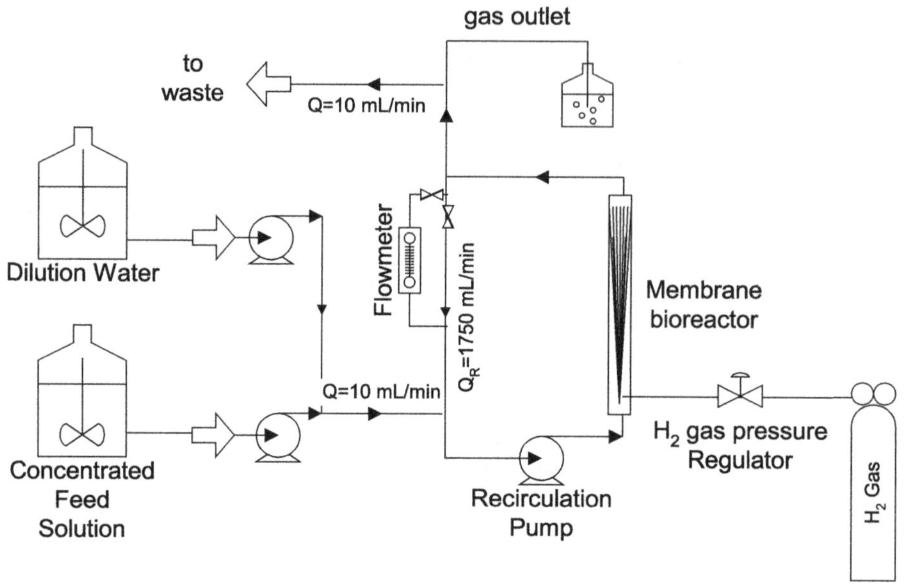

Figure 3.1 Schematic of the intial studies membrane biofilm reactor (MBfR)

Table 3.1
Characteristics of the hollow-fiber membrane biofilm reactor system

Item	Value	Unit
Pipe Length	120	cm
Pipe Internal Diameter	1.544	cm
Pipe Cross Sectional Area	1.872	cm^2
Fiber Outside Diameter	280.0	μm
Fiber Cross Sectional Area	0.00062	cm^2
Number of Fibers	98	
Length of Fibers	92.7	cm
Active Length of Fibers	72.1	cm
Area of Single Fiber	0.06	cm^2
Surface Area of Fibers	624	cm^2
Feed Flow Rate	10	mL/min
Detention Time	44	min
Recycle Flow	1750	mL/min
Recycle Ratio	175	

Table 3.2
Typical tap water quality parameters

Parameter	Typical concentration (mg/L)
ClO_4^-	1
Cl^-	12.4
NO_3^- (as N)	0.55
NO_2^- (as N)	<0.1
$SO_4^=$	23.2
pH	7
Hardness (as $CaCO_3$)	138
Alkalinity (as $CaCO_3$)	103

Feed Solutions

For these screening and mechanisms experiments, the reactor was supplied with concentrated feed solution mixed with oxygen-free dilution water. Oxygen was removed from the dilution water by continuously sparging with helium or argon. In the screening experiments, tap water was used for dilution, while reverse osmosis treated water was used for dilution in the mechanisms experiments. In the groundwater experiments, no dilution water was required, and oxygen was not removed from the influent. In all experiments, the feed rate was 10 mL/min, the hydraulic detention time was 44 minutes, and water was recirculated within the reactor at a rate of 1,750 mL/min. In addition to fluidizing the fibers, the high recycle flow rate helped promote completely mixed conditions and control biomass accumulation on the fibers.

For the screening experiments, tap water was supplemented with 1 mg/L perchlorate and 5 mg-N/L nitrate, and the influent pH was 7.0 (See Table 3.2). Initially, excess hydrogen was achieved by maintaining a 7.0-psi (0.48-atm) hydrogen supply pressure to the hollow fiber membrane. Subsequently, the pressure was dropped to 3.0-psi (0.20-atm), which was adequate for nitrate and perchlorate reduction.

The mechanism experiments used a minimal medium, shown in Table 3.3, which Aragno and Schlegel (1992) employed for autotrophic hydrogen-oxidizing, denitrifying bacteria. The influent pH was 7 for the mechanism experiments. The influent contained 1 mg/L perchlorate and 5 mg-N/L nitrate. The hydrogen pressure applied to the fiber was 2.5-psi (0.17-atm).

Water for the groundwater experiment was collected from a well owned and operated by the Suburban Water Company, located in the Main San Gabriel Basin, California, and periodically shipped in coolers. Once received, the water was stored in a cold room at 4°C and brought to room temperature (20°C) before feeding to the reactor. No chemical additions were made during the groundwater experiments other than perchlorate spikes and hydrogen delivered by the MBfR. The quality of the groundwater sample is shown in Table 3.4. Groundwater tests with the hydrogen reactor were performed previously at steady state with 1-mg/L perchlorate and 5-mg-N/L nitrate. Three phases of testing spanned a 28-day period. First, the groundwater was applied for 7 days with its natural perchlorate concentration of 6 μg/L to represent a low-perchlorate scenario.

Table 3.3
Minimal medium

Component	Concentration (mg/L)
$NaNO_3$	30.36
Na_2HPO_4	142
KH_2PO_4	136
$NaHCO_3$	5.1
$NaClO_4$	1.225
$MgSO_4 \cdot 7H_2O$	200
$CaCl_2 \cdot 2H_2O$	1
$FeSO_4 \cdot 7H_2O$	1
EDTA	3
$ZnSO_4 \cdot 7H_2O$	1
$MnCl_2 \cdot 4H_2O$	0.3
H_3BO_3	3
$CoCl_2 \cdot 6H_2O$	0.2
$CuCl_2 \cdot 2H_2O$	0.1
$NiCl_2 \cdot 6H_2O$	0.2
$Na_2MoO_4 \cdot 2H_2O$	0.3

Table 3.4
Quality of California groundwater sample

Parameter	Unit	Value
Perchlorate	μg/L	6
Nitrate	mg-N/L	2.5-3
pH	—	7.7
Alkalinity	mg/L as $CaCO_3$	202
TOC	mg/L	1.4

Second, the feed water was spiked with 100-μg/L perchlorate for 15 days, to represent a high range of perchlorate. Finally, the reactor feed was spiked with 50 μg/L for 5 days, representing a mid-range of perchlorate contamination. The reactor operating conditions were similar to those used in the previous experiments, except the applied hydrogen pressure was 5 psi (0.34 atm).

Analytical Methods

Most of the parameters were measured following *Standard Methods* (APHA, AWWA, and WEF, 1995) or USEPA Method 300.1 (Hauman and Munch, 1997). Analyses following *Standard Methods* included chloride (Cl^-), nitrate (NO_3^-), nitrite (NO_2^-), total organic carbon/dissolved organic carbon (TOC/DOC), dissolved oxygen (DO), pH, and alkalinity. Analyses following USEPA Method 300.1 (Hauman and Munch, 1997) included ClO_3^- and ClO_2^-. Analytical methods used are summarized in Table 3.5.

Low-level perchlorate was analyzed using an ion chromatograph (IC) (Dionex 4000i, Dionex, Sunnyvale, CA) with conductivity detection and an AS-16 column, a 500-mL loop, and an autosampler. Based on repeated injections of a 10-μg/L standard, the MRL was 2 μg/L. The IC was used with an AS-11 column to analyze for nitrate, nitrite, chlorate, chlorite, and chloride. The AS-11 column was also used to determine perchlorate in some experiments with high perchlorate concentrations.

Dissolved hydrogen was analyzed via headspace disproportionation (Schmidt and Ahring, 1993). For this analysis, a 1-ml liquid sample was transferred from the reactor to a 160-ml serum vial previously outgassed with nitrogen. The vial was shaken vigorously to liberate dissolved hydrogen. A gas-tight syringe was used to sample the headspace (1 ml) and test for hydrogen by reduction gas analysis (RGA3, Trace Analytical, Menlo Park, CA). In this method, hydrogen is directed through a HgO bed and produces Hg(g), which is measured by an ultraviolet photometer. Once the hydrogen concentration was known, Henry's law and mass balance were used to determine the dissolved hydrogen concentration.

RESULTS AND DISCUSSION

Screening Tests

The purpose of the screening experiment was to determine if hydrogen was an effective electron donor for perchlorate reduction in the MBfR. The experiment commenced when approximately 1 mg/L perchlorate was first introduced to a 5 mg-N/L nitrate feed stream. Nitrate removal (not shown) was 96 percent after one day, increasing to 99 percent after the 18th day. Perchlorate removal (Figure 3.2) was 39 percent after one day, 90 percent after 7 days, and 99 percent after 18 days. No accumulation of intermediates (i.e., ClO_3^-, ClO_2^-) was detected. The chloride concentration increased in approximately stoichiometric amounts, although exact measurements were not possible due to the relatively high background chloride concentration in the tap water. The lowest measured effluent perchlorate concentration was 13 μg/L.

The screening experiments verified that hydrogen was an effective donor, and that no special inoculation was required. The MBfR initially removed approximately 30 percent of the perchlorate, even though the biofilm had not been previously exposed to perchlorate and no specific PCRB seed had been introduced. Some form of adaptation occurred for the perchlorate removal to increase to 99 percent over the three-week screening experiment. The most likely

Table 3.5
Summary of analytical procedures

Parameter	Method number	Method title	Reference
ClO_4^-	Modified EPA Method 300.1	Analysis of Low Concentrations of Perchlorate in Drinking Water by Ion Chromatography	Dionex Application Note 121
ClO_3^-	EPA Method 300.1		Hauman and Munch (1997)
ClO_2^-	EPA Method 300.1		Hauman and Munch (1997)
Cl^-	4500-Cl^- F	Chloride-Ion Chromatography Method	*Standard Methods* (APHA, AWWA, and WEF, 1995)
$H_{2\,(g)}$		Headspace	Schmidt and Ahring (1993)
NO_3^-	4500-NO_3-C	Ion Chromatographic Method	*Standard Methods* (APHA, AWWA, and WEF, 1995)
NO_2^-	4500-NO_2-C	Ion Chromatographic Method	*Standard Methods* (APHA, AWWA, and WEF, 1995)
TOC/DOC	5310 B	Combustion Infrared	*Standard Methods* (APHA, AWWA, and WEF, 1995)
DO	4500-O G	Membrane Electrode Method	*Standard Methods* (APHA, AWWA, and WEF, 1995)
pH	4500-H^+	Electrometric Method	*Standard Methods* (APHA, AWWA, and WEF, 1995)
Conductivity	2510	Electrical Conductivity	*Standard Methods* (APHA, AWWA, and WEF, 1995)
Alkalinity	2320 B	Titration Method	*Standard Methods* (APHA, AWWA, and WEF, 1995)

cause for the adaptation to perchlorate was an enrichment of PCRB. A PCRB species capable of using nitrate and perchlorate would have had a metabolic advantage over common denitrifiers. Since the perchlorate loading was less than 5 percent of that of nitrate on a hydrogen-accepting basis, the number of PCRB may have been small compared to the common denitrifiers.

Perchlorate and nitrate fluxes were determined from this experiment. For perchlorate, with an influent concentration of 1.14 mg/L, an effluent of 0.013 mg/L, a flow of 10 mL/min, and a

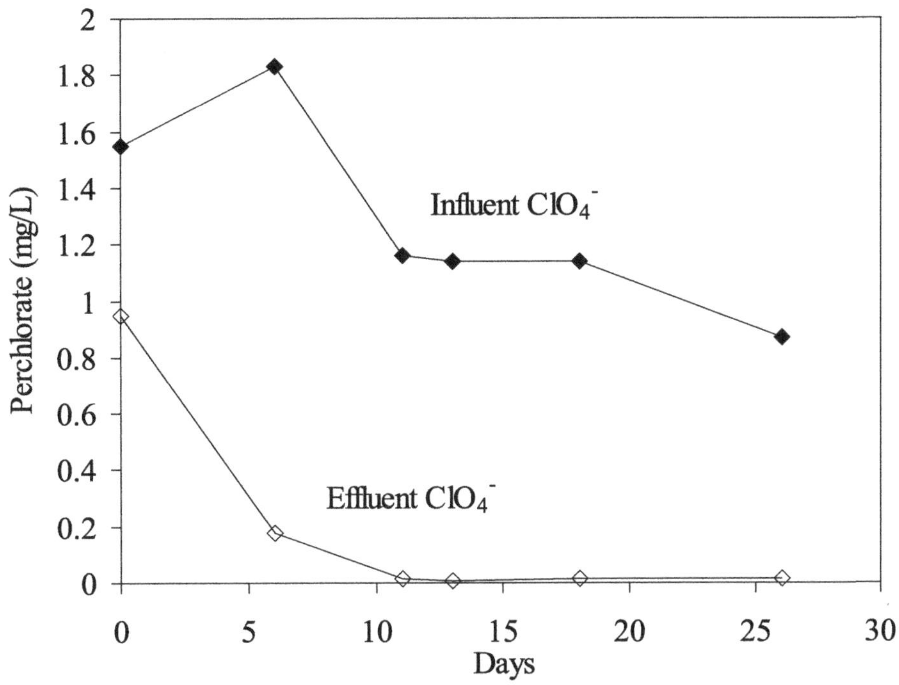

Figure 3.2 Fate of perchlorate in the screening experiment

fiber surface area of 624 cm^2, the perchlorate flux is 260 $mgClO_4^-/m^2$-day. For nitrate, with 5.8 mgN/L influent and 0.052 mgN/L effluent, the flux is 1326 mgN/m^2-day.

Mechanisms Experiments

Experiments were conducted to explore if and how the hydrogen, perchlorate, and nitrate concentrations and pH affected perchlorate reduction. Once the biofilm in the MBfR reached steady state, pseudo-steady-state conditions were achieved by varying a single parameter during a time frame long enough to reach hydraulic steady-state, but short enough to prevent appreciable changes in biomass. Prior to the experiments, the reactor was cleaned and allowed to return to steady-state conditions with an influent with 1 mg/L perchlorate, 5 mg-N/L nitrate, 2.5 psi (0.17 atm) hydrogen pressure, and an RO water based minimal medium.

Hydrogen Variable

In the first short-term experiment, the applied hydrogen pressure was varied among 1.5 psi (0.10 atm), 2.5 psi (0.17 atm), 4 psi (0.27 atm), and 5.5 psi (0.37 atm). Results are shown in Figure 3.3. At 1.5 psi (0.10 atm), the effluent residual hydrogen was 11 μg/L, providing partial reduction of nitrate to 0.03 mg-N/L, or 86 percent removal. Under these conditions, the perchlorate reduction was only 5 percent. At 2.5 psi (0.17 atm) applied pressure, the effluent residual hydrogen was 37 μg/L, and denitrification reached its maximum removal of 99 percent. Nitrate removals did not increase further at higher hydrogen pressures. At 2.5 psi, perchlorate reduction increased to 15 percent. At an applied pressure of 4 psi, with effluent residual hydrogen of 310 μg/L, perchlorate reached its maximum removal of 35 percent.

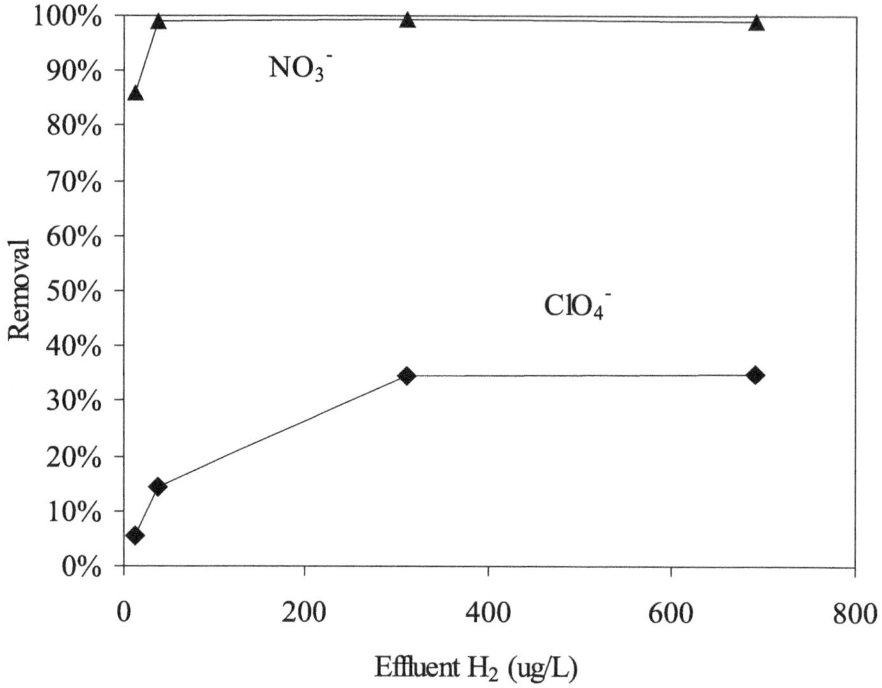

Figure 3.3 Mechanisms experiment 1—hydrogen variable

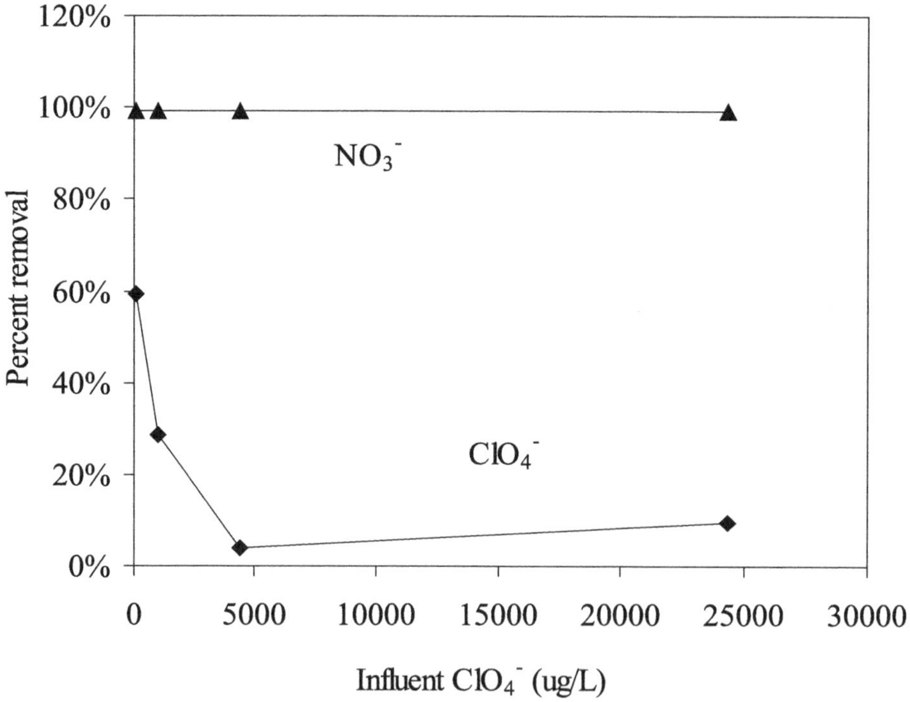

Figure 3.4 Mechanisms experiment 2—perchlorate variable

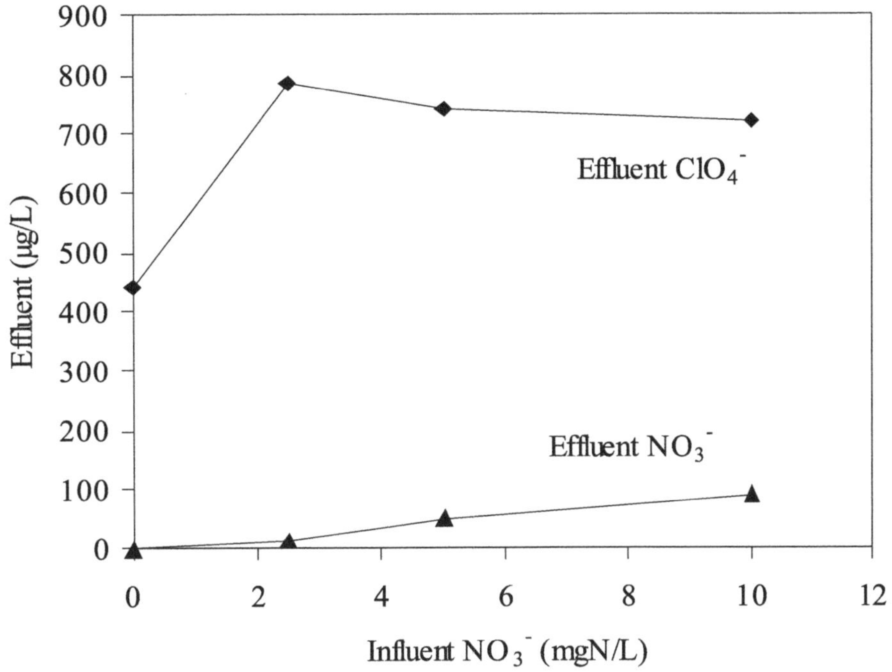

Figure 3.5 Mechanisms experiment 3—nitrate variable

Perchlorate Variable

In the second short-term experiment, the influent perchlorate concentration was varied among 0, 0.2, 1, 5, and 25 mg/L, with applied hydrogen and influent nitrate at the steady-state values of 2.5 psi (0.17 atm) and 5 mg-N/L, respectively. Results are shown in Figure 3.4. Although nitrate reduction was not affected by the influent perchlorate load, the percent perchlorate removal decreased significantly with the increased perchlorate load. Chlorite was not detected in any of the experiments. However, a small amount of chlorate, 0.33 mg/L, was detected during the experiment with 25-mg/L perchlorate (data not shown). The effluent hydrogen concentration varied from 30 to 50 µg/L during these experiments (data not shown).

Nitrate Variable

In the third short-term experiment, the influent nitrate concentrations were varied from 0 to 15 mg-N/L, with hydrogen and perchlorate at the steady-state levels of 2.5 psi (0.17 atm) and 1 mg/L, respectively. Results are shown in Figure 3.5. With zero nitrate in the influent, no nitrate was found in the effluent, and perchlorate reduction increased to 57 percent. However, with influent nitrate at 2.5 mg-N/L or higher and the effluent nitrate concentrations at or above 0.014 mg-N/L, perchlorate reduction decreased to 25 to 30 percent. The hydrogen concentration in the effluent was inversely proportional to the nitrate loading.

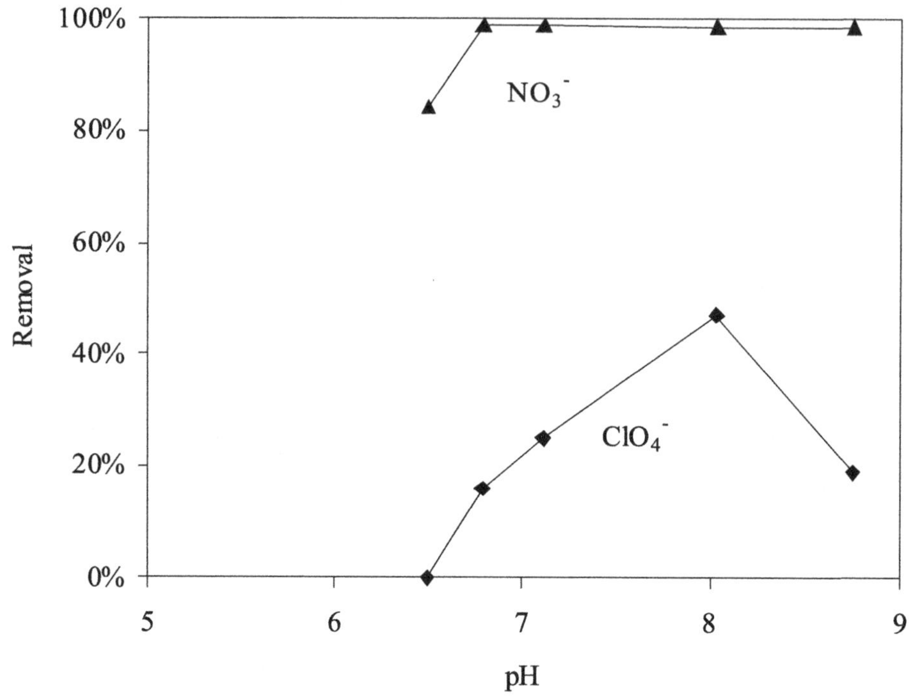

Figure 3.6 Mechanisms experiment 4—effluent pH variable

pH Variable

In the fourth short-term experiment, the influent pH was set to 6.5, 6.8, 8.03, and 8.75, with applied hydrogen and influent nitrate at their steady-state values of 2.5 psi (0.17 atm) and 5 mgN/L, respectively. Results are shown in Figure 3.6. Although nitrate reduction was hardly sensitive to pH, perchlorate reduction was pH sensitive, and the best removal occurred at a pH of 8.

The mechanisms experiments showed that the maximum perchlorate reduction required a minimum hydrogen residual of about 300 μg/L in the bulk liquid, while maximum nitrate reduction only required 50 μg/L. Furthermore, perchlorate reduction was very sensitive to the nitrate concentration: Reduction was 57 percent with zero nitrate, but dropped to about 30 percent with nitrate at levels as low as 0.014 mg-N/L. Nitrate inhibition of perchlorate reduction, however, did not increase as the effluent nitrate concentration increased from 0.014 to 0.23 mg/L. The observation of inhibition may indicate that perchlorate reduction is incompatible with partial denitrification (Lee and Rittmann, 2000). On the other hand, partial inhibition of perchlorate reduction by nitrate suggests either that some PCRB within the biofilm were not subject to nitrate inhibition, such as HAP-1 (Wallace et al., 1996), or that nitrate did not penetrate into the deeper portions of biofilm, where perchlorate reduction occurred free of nitrate inhibition.

The optimal pH for perchlorate reduction was around 8, and the range for perchlorate reduction appeared to be from 6.8 to around 9. The lack of perchlorate reduction at pH 6.5 may have been exacerbated by nitrate inhibition, since nitrate reduction was partially inhibited at this pH. The sensitive pH effects on perchlorate reduction are significant, because denitrification adds base, which can cause a pH increase that might slow perchlorate reduction and limit the accumulation of PCRB.

An additional mechanism experiment was performed to determine the fate of chlorite in the reactor. With an influent containing 5 mg-N/L nitrate, 10 mg/L chlorite, and no perchlorate, over 95 percent of the chlorite was removed, with over 80 percent transformed to chloride. A trace amount of chlorate, 0.17 mg/L, was detected. With an influent containing 5 mg-N/L nitrate, 1 mg/L chlorite and no perchlorate, neither chlorite nor chlorate was detected. Chlorate accumulation has not been reported for pure cultures of PCRB.

The chlorite-addition test confirmed that chlorate can be produced from chlorite in the reactor. More importantly, these experiments confirmed that chlorite dismutation to chloride and oxygen is faster than perchlorate reduction and that intermediate accumulations normally do not occur.

Groundwater Experiments

As the previous tests were all conducted with amended tap water or RO-based minimal media, an additional experiment was performed to test the reactor with a groundwater from a contaminated site and without chemical amendments. The test included phases with different levels of perchlorate in the influent: 6 μg/L, 100 μg/L, and 50 μg/L. The influent nitrate concentration was 2.6 to 3.0 mg-N/L. The influent and effluent perchlorate results are shown in Figure 3.7. For all three phases of the test, the effluent perchlorate varied from non-detect (less than 2 μg/L) to 4.5 μg/L, and the effluent nitrate varied from 13 to 32 μgN/L, which corresponds to at least 99 percent removals of perchlorate and nitrate.

The flux for perchlorate, based on an influent of 0.105 mg/L, an effluent of 0.004 mg/L, a flow of 10 mL/min, and a fiber surface area of 624 cm^2, is 23 $mgClO_4^-/m_2$-day. For nitrate, based on an influent nitrate concentration of 2.57 mgN/L and an effluent of 0.025 mgN/L, the flux is 587 mgN/m_2-day.

The groundwater experiments clearly demonstrated the feasibility of using the MBfR to remove perchlorate from groundwater. The bench-scale reactor consistently achieved removals at or below the MRL of 4 μg/L over a four-week period. The reactor also responded well to sudden changes in perchlorate concentrations. For example, when the influent concentration was suddenly increased from 6 μg/L to 100 μg/L, the effluent perchlorate concentrations did not increase above the 4 μg/L MRL.

Hydrogen Reactor Minimal Medium Investigation

The hydrogen reactor's steady-state perchlorate removals during the mechanisms characterization were substantially lower (i.e., 30%) than those of the initial feasibility screening (95-99%). Poorer steady-state removal of perchlorate was a goal of the experiments, as it allowed increases and decreases in perchlorate removal to be observed. However, after completing the mechanisms experiments and maintaining the reactor under the same steady-state conditions for several months, the perchlorate reduction progressively deteriorated until it reached less than 10 percent. Meanwhile, nitrate removal efficiencies continued at around 99 percent.

The loss of perchlorate reduction appeared to be related to the medium used in the mechanisms experiments. Subsequent experiments were performed to ascertain whether or not the tap water was responsible for greater perchlorate removal. Switching between the RO-based minimal media and the amended tap water, as shown in Figure 3.8, revealed that the minimal media lacked something essential for perchlorate reduction, but not for denitrification.

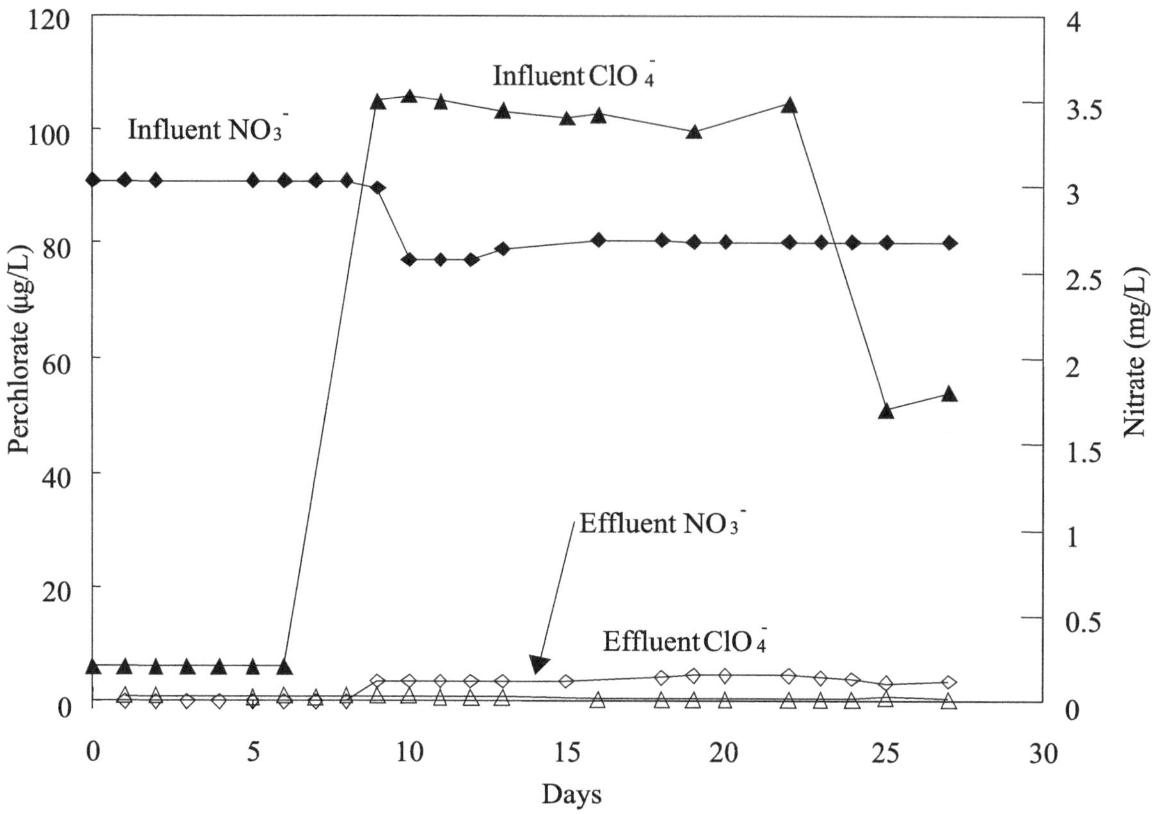

Figure 3.7 Perchlorate concentrations in groundwater experiment

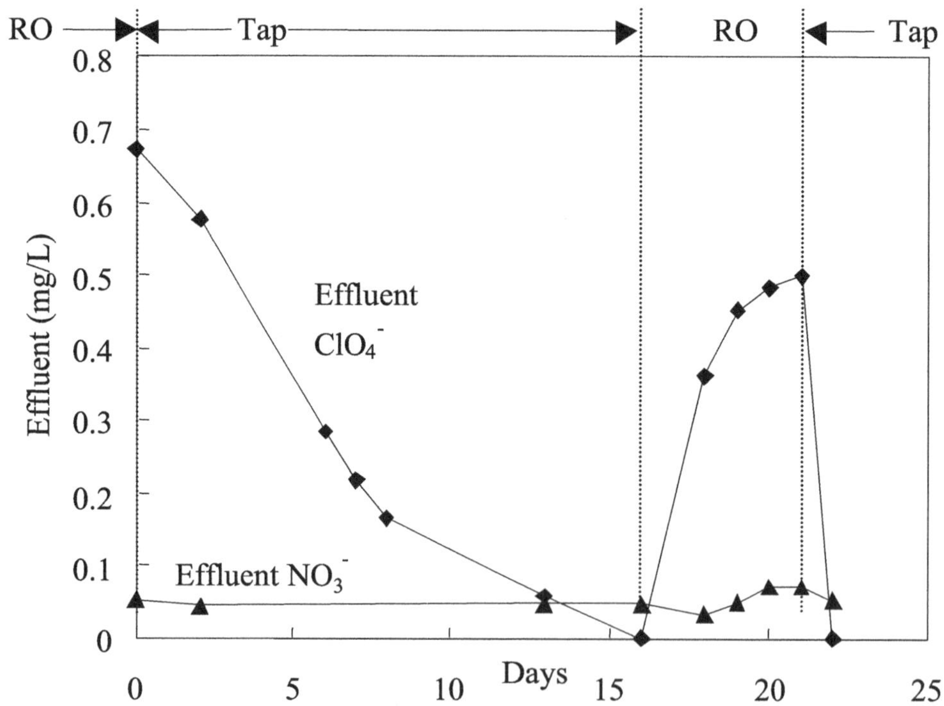

Figure 3.8 Effects of medium on MBfR performance

The results also suggested that the reactor's microbial ecology plays an important role in perchlorate removal efficiency. After continued exposure to the RO media, perchlorate reduction was nearly lost, and it took several weeks to recover it when tap water and perchlorate were present. However, after only a short exposure to the RO media, recovery of perchlorate reduction was nearly immediate. The tap water medium with perchlorate may enrich for PCRB, and the RO medium may cause them to gradually be lost from the biofilm.

INITIAL FEASIBILITY INVESTIGATIONS CONCLUSIONS

The initial investigations clearly demonstrated that biological reduction of perchlorate to chloride is technically feasible in the drinking-water setting using hydrogen as an electron donor. For example, the hydrogen-based MBfR successfully reduced perchlorate concentrations below the 4-μg/L MRL when the influent perchlorate concentration was as high as 100 μg/L in an actual contaminated groundwater. Additionally, perchlorate reduction in the MBfR occurred immediately upon exposure to perchlorate and without inoculation with specialized PCRB.

The mechanisms studies showed that excellent perchlorate removal was possible when only a small residual concentration of the electron donor (300 μg/L of hydrogen) was present in the effluent. However, nitrate in the water slowed perchlorate reduction, suggesting that partial nitrate removal may not be feasible when perchlorate reduction is a treatment goal. Conversely, the presence of perchlorate had no impact on denitrification. Finally, natural surface water and groundwater contained unknown factors that selected for PCRB better than a minimal medium made in RO water.

Perchlorate reduction does not appear to be linked to co-metabolic reduction by denitrifying bacteria. Rather, it seems to be linked to the presence of specialized bacteria, probably PCRB, which can be lost when the conditions are not suitable.

CHAPTER 4
PERCHLORATE REDUCTION KINETICS

In Chapter 3, it was shown that perchlorate could be reduced to below 4 μg/L in a hydrogen-fed membrane biofilm reactor (MBfR). In order to better understand the fundamentals of perchlorate reduction under autotrophic, hydrogen-oxidizing conditions, kinetic parameters were determined for a pure culture. *Dechloromonas* sp. PC1, a novel, autotrophic hydrogen-oxidizing PCRB isolated from the bench-scale MBfR used in the Chapter 3 was employed. Inherent to the evaluation of kinetic parameters was evaluating and quantifying the effects of competitive inhibition, as described below.

MATERIALS AND METHODS

Chemicals and Medium

The growth medium, adapted from Aragno and Schlegel (1993), was prepared using ultra-pure water (Nanopure, Barnstead/Thermolyne, Dubuque, Iowa) and research-grade chemicals. The growth medium contained, per liter: 1.386 g Na_2HPO_4, 0.849 g KH_2PO_4, 0.1 g $(NH_4)_2SO_4$, 0.2 g $MgSO_4 \cdot 7H_2O$, 1 mL trace mineral solution, and 1 mL Ca-Fe Solution. The Ca-Fe Solution contained, per liter: 1 g $CaCl_2 \cdot 2H_2O$ and 1 g $FeSO_4 \cdot 7H_2O$. The trace mineral solution contained, per liter: 100 mg $ZnSO_4 \cdot 7H_2O$, 30 mg $MnCl_2 \cdot 4H_2O$, 300 mg H_3BO_3, 200 mg $CoCl_2 \cdot 6H_2O$, 10 mg $CuCl_2 \cdot 2H_2O$, 10 mg $NiCl_2 \cdot 6H_2O$, 30 mg $Na_2MoO_4 \cdot 2H_2O$, and 30 mg Na_2SeO_3. The pH was adjusted using 1 M NaOH, for a final pH of 7.0, after degassing and adding a gas mixture with 95% H_2 plus 5% CO_2. The non-growth media was a 4-mM phosphate buffer at pH 7. Agar plates were prepared with R2A agar (DIFCO, Detroit, MI).

Analytical Methods

Perchlorate, nitrate, nitrite, chlorate, chlorite, and chloride were analyzed using Dionex 4000i (Sunnyvale, CA) ion chromatograph with a 4-mm AS-11 column, a GS-11 pre-column, and a conductivity detector. A sodium hydroxide eluant was used, with a gradient from 4 mM to 60 mM. A suppressor (Dionex ASRS) was used in chemical suppression mode with a 75 mM H_2SO_4 suppressant at 5 mL/min. Samples were injected with an autosampler (Dionex).

Suspended biomass was determined by correlation with optical density (OD). The OD was measured at 600 nm with a spectrophotometer (Spec 20, Thermo Spectronics, Rochester, NY), and converted to dry weight using an empirical conversion factor determined for PC1. The conversion factor was determined by growing bacteria to exponential phase on perchlorate to an optical density of around 0.2, filtering a known volume on a 0.45-μm cellulose filter (Gelman Sciences, Ann Arbor, MI), drying the filter to constant weight at 104°C, and measuring the net dry weight, i.e., the dry weight of the sample minus the dry weight of the filter paper. The conversion factor was 588 mgDW/OD unit. The linearity of the calibration was confirmed by measuring serial dilutions of a sample of known OD and plotting the results.

Kinetics and Inhibition Experiments

Batch tests were used to determine kinetic parameters and assess competitive inhibition for strain PC1. The true yield (Y) and maximum specific substrate utilization rate (q_{max}) were determined with batch-growth experiments using high initial perchlorate and low initial biomass concentrations, while the half-maximum-rate concentration (K) was determined with non-growth experiments using low initial biomass and perchlorate concentrations. The endogenous decay rate, b, was determined by growing cells to high densities with hydrogen as an electron donor, purging the hydrogen with oxygen-free nitrogen, then measuring the decrease in optical density with time. Tests were carried out in 1-L bottles filled with 200 mL of media or 160-mL serum bottles filled with 25 mL of sterile media. All bottles were capped with butyl rubber stoppers, vacuum degassed, and filled with a gas mixture of 95% hydrogen and 5% CO_2 (q_{max} and Y) or with pure hydrogen (K). The bottles were shaken on their sides at 200 rpm. The experiments were carried out at least in triplicate. The pH was 7, and the temperature was 22°C.

Organism Isolation and 16S rDNA Identification

Strain PC1 was isolated from the MBfR used in the initial investigations by adding effluent from the reactor to 160-mL serum bottles containing 25 mL of growth medium with perchlorate. An autotrophic, hydrogen-oxidizing PCRB was isolated from the enrichment by plating aerobically with R2A agar (DIFCO, Detroit, MI) and then re-growing in liquid medium. Purity was confirmed by plating and by sequencing. For the testing electron acceptors, 160-mL serum bottles with the growth medium described above were used. 1 mM of the tested acceptor was added and growth was measured by increases in absorbance over a 3-week period.

Scanning Electron Microscopy (SEM)

For SEM, an anaerobically grown pure-culture suspension of PC1 was filtered onto a 13-mm diameter, 0.45-μm pore size cellulose filter and immersed into a 1% gluteraldehyde solution. After washing in a 16-mM phosphate buffer at pH 7, the sample was dehydrated by immersing in ethanol baths at 10%, 30%, 50%, 70%, 90%, and twice in a bath at 100%. The sample was critical-point dried (Polaron Critical Point Drier, Quorum Technologies, East Sussex, England) and sputter-coated the filter with gold to a 3-nanometer thickness (Cressington 208HR sputter coater, Cranberry Twp., PA). Subsequently, the sample was viewed on a scanning electron microscope (Hitachi 4500, Pleasanton, CA).

Numerical Methods

Non-linear curve fitting was used to estimate kinetic parameters q_{max} and K based on a finite-differences solution of the substrate-utilization and biomass-growth equations. When chlorate or nitrate were the added acceptor, q_{max} and Y was determined by fitting the standard Monod substrate-utilization and biomass-accumulation equations for batch growth:

Chlorate or nitrate mass balance

$$\frac{dS}{dt} = -\frac{q_{\max}S}{S+K}X \tag{4.1}$$

Biomass mass balance

$$\frac{dX}{dt} = \frac{Yq_{\max}S}{S+K}X - bX \tag{4.2}$$

where

Y is the biomass true yield [M_X/M_S] under autotrophic conditions with hydrogen as a donor

b is the endogenous decay rate [1/T]

q_{max} is the maximum specific substrate utilization rate [M_X/M_S-T]

S is the nitrate or chlorate concentration [M_S/V]

K is the half-maximum substrate-utilization rate constant [M_S/V] for nitrate or chlorate

The standard unit system is days for time [T], mg dry weight for biomass [X], mg chlorate or nitrate for substrate [S], and liters for volume [V].

Equations 4.1 and 4.2 were used to determine kinetic parameters for perchlorate, even though chlorate accumulation slows the perchlorate reduction rate. This was possible because, when the perchlorate concentrations are sufficiently high, the apparent reduction rate with inhibition, q_P, is constant. It is related to q_{maxP} through the following equation:

$$q_{\max P} = \frac{q_P q_{\max C}}{q_{\max C} - q_P} \tag{4.3}$$

Thus, by measuring q_P in batch tests and by knowing q_{maxC} from other experiments, q_{maxP} can be determined. A different model was used to determine the K_C for chlorate and to assess the competitive inhibition model.

Since perchlorate and chlorate are reduced by the same enzyme, (per)chlorate reductase, they presumably inhibit each other competitively (Engel, 1981; Rittmann and Sáez, 1993), meaning that the presence of one slows the reduction rate of the other. Since perchlorate reduction produces chlorate, any accumulating chlorate may slow perchlorate reduction rates. High perchlorate concentrations also could promote chlorate accumulation by competing with chlorate and slowing its reduction rate. In order to determine the true kinetics of perchlorate reduction, i.e., those that are independent of the chlorate concentration, competitive inhibition was considered in the modeling equations.

In this case, perchlorate reduction was modeled in two steps: reduction of perchlorate to chlorate, and reduction of chlorate to chloride. This approach is reasonable because the disproportionation of chlorite to oxygen and chloride and the reduction of oxygen to water are very fast compared to reduction of chlorate (van Ginkel et al., 1996; Logan et al., 2001).

Similarly, biomass yields were broken down into yields from reduction of perchlorate to chlorate and reduction of chlorate to chloride. The non-steady-state mass balance and rate expressions are:

Perchlorate mass balance

$$\frac{dS_P}{dt} = -q_P X \tag{4.4}$$

Chlorate mass balance

$$\frac{dS_C}{dt} = (\alpha_{C/P} q_P - q_C) X \tag{4.5}$$

Biomass mass balance

$$\frac{dX}{dt} = (Y_P q_P + Y_C q_C - b) X \tag{4.6}$$

Perchlorate reduction rate

$$q_P = q_{P\max} \frac{S_P}{S_P + K_P\left(1 + \frac{S_C}{K_C}\right)} \tag{4.7}$$

Chlorate reduction rate

$$q_C = q_{C\max} \frac{S_C}{S_C + K_C\left(1 + \frac{S_P}{K_P}\right)} \tag{4.8}$$

where

S_P and S_C are the perchlorate and chlorate concentrations $[M_S/V]$
X is the biomass concentration $[M_X/V]$, q_P is the effective specific perchlorate reduction rate $[M_S/M_X\text{-}T]$
$q_{P\max}$ is the maximum specific perchlorate reduction rate $[M_S/M_X\text{-}T]$
q_C is the effective specific perchlorate reduction rate $[M_S/M_X\text{-}T]$
$q_{C\max}$ is the maximum specific chlorate utilization rate $[M_S/M_X\text{-}T]$
Y_P is the yield obtained from reducing perchlorate to chlorate $[M_X/M_S]$
Y_C is the yield from reducing chlorate to chloride and oxygen to water $[M_X/M_S]$
K_P and K_C are the half-maximum-rate concentrations for perchlorate and chlorate [M/V]
$\alpha_{C/P}$ is the conversion factor between the mass of perchlorate reduced and the mass of chlorate produced by perchlorate reduction $[M_{SC}/M_{SP}]$.

Competitive inhibition in the rate expressions for q_C and q_P (Equations 4.7 and 4.8) is governed by the term in parentheses. In Equation 4.7, when the inhibiting acceptor S_C is present at concentrations that are high relative to K_C, the parenthesis term is greater than 1, in effect increasing the value of the Monod K term, which in turn decreases the reduction rate q_P. When the inhibiting acceptor, S_C, is low with respect to K_C, the inhibiting effect is small. The competitive inhibition kinetic expressions (Equations 4.6 and 4.7) are theoretically derived from the expressions for competitive inhibition in free enzymes (Engel, 1981).

Curve-fitting was carried out using relative least-squares minimization (Sáez and Rittmann, 1992). In this technique, the modeling equations are solved numerically, and parameters are selected to minimize the sum of the relative least-square residuals, as shown in Equation 4.9.

$$F = \sum_{i=1}^{n} \left[\frac{y_i - y_i^*}{y_i} \right]^2 \qquad (4.9)$$

where

n is the number of experimental observations
y_i is the measured value for observation number I
y_i^* is the model-predicted value for observation number i.

The advantage of relative least-squares over an absolute least-square method is that it eliminates bias from the high errors associated with higher concentrations.

For chlorate, perchlorate, and nitrate, Y and q_{max} were determined concurrently by minimizing the sum of the relative least squares for substrate and biomass. The b and K parameters were determined in separate experiments with low initial biomass and substrate concentrations. For chlorate, additional experiments were required to determine K, as discussed below. For all parameters, curve-fitting was performed by solving the differential equations on a Microsoft Excel spreadsheet using a finite-differences scheme and performed the minimization of relative least squares using the Excel "Solver" optimization routine. In order to determine the parameter K_C, the K_C was fit to experimental data with either perchlorate reduction or concurrent perchlorate and chlorate reduction.

RESULTS AND DISCUSSION

16S rDNA Identification and Physiology

Strain PC1 is within the genus *Dechloromonas* of the *Rhodocyclus* group of the *beta*-subclass of the *Proteobacteria*. The *Dechloromonas* genus is one of two that include most known species of perchlorate-reducing bacteria. A search for similar nucleotide sequences from BLAST revealed that PC1 had a 99-percent similarity to *Dechloromonas* strain JJ (Coates et al., 2001), which is not a PCRB, and a 98-percent similarity to *Dechloromonas* strain HZ, which is a hydrogen-oxidizing, perchlorate-reducing autotroph (Zhang, Bruns, and Logan, 2002).

PC1 is a Gram-negative motile rod with a 1-µm average length and 0.3-µm average diameter. Figure 4.1 shows a scanning electron micrograph of PC1. PC1 is a facultative anaerobe and can grow anaerobically with perchlorate, chlorate, nitrate, and nitrite as electron acceptors. It

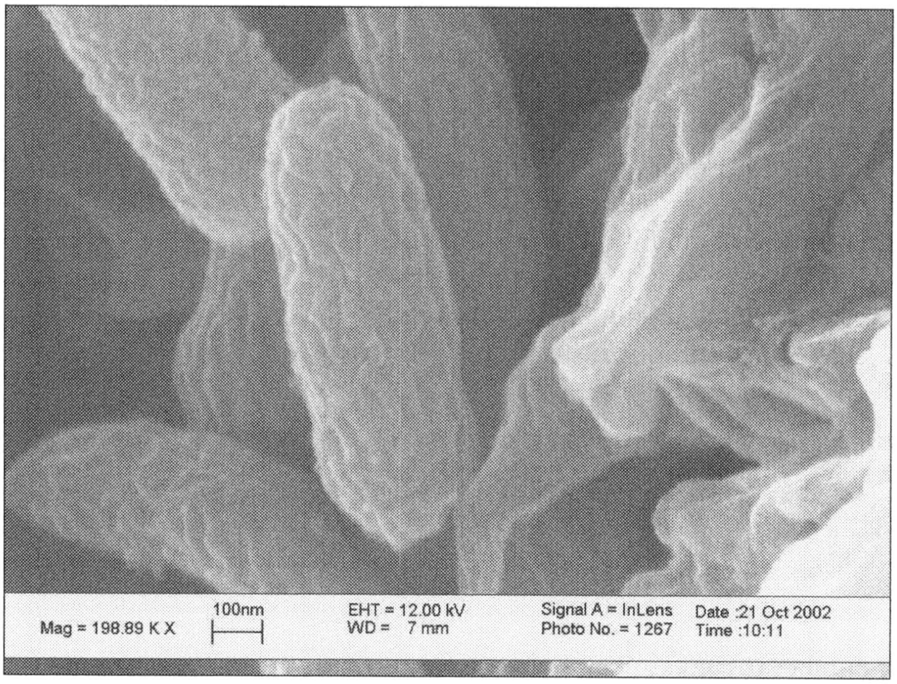

Figure 4.1 SEM image of isolate PC1

cannot grow on sulfate, bromate, selenate, selenite, sulfate, or arsenate. It can grow autotrophically with hydrogen as an electron donor, and heterotrophically with acetate.

Kinetic Parameters

Chlorate

Typical growth curves and K tests for *Dechloromonas* sp. PC1 for chlorate are shown in Figure 4.2 and Figure 4.3. The true yield for chlorate is 0.22 gDW/gClO_3^-. The yield, expressed in terms of biomass per electron transferred to the acceptor, is 2.8-gDW/eq e^-. The McCarty method (Rittmann and McCarty, 2001) allows estimation of microbial yields based on thermodynamics of the donor, acceptor, carbon and nitrogen sources, bacterial maintenance requirements, and growth conditions. For autotrophic growth with hydrogen as an electron donor and ammonium as a nitrogen source, the expected true yield for chlorate is 2.64 gDW/eq e^-. Thus, the determined yield for PC1 is consistent with the theoretical value. The q_{max} value for chlorate is 6.3 gClO_3^-/gDW-day. On the basis of electrons transferred to the acceptor, the q_{max} is 0.31 e^-eq/gDW-day.

The K for chlorate is less than 2 μg/L, which was too low to determine it from the non-growth batch tests (Figure 4.3). However, K_C was determined indirectly through its inhibiting affect on perchlorate reduction (Equation 4.7) and found it to be 0.0012 mg/L. This is discussed later in this chapter.

Perchlorate

The growth and K experiments for perchlorate are shown in Figure 4.4 and Figure 4.5. The total yield on perchlorate is 0.23 gDW/gClO_4^-, or 2.9 gDW/eq e^- on a per-electron basis. This is

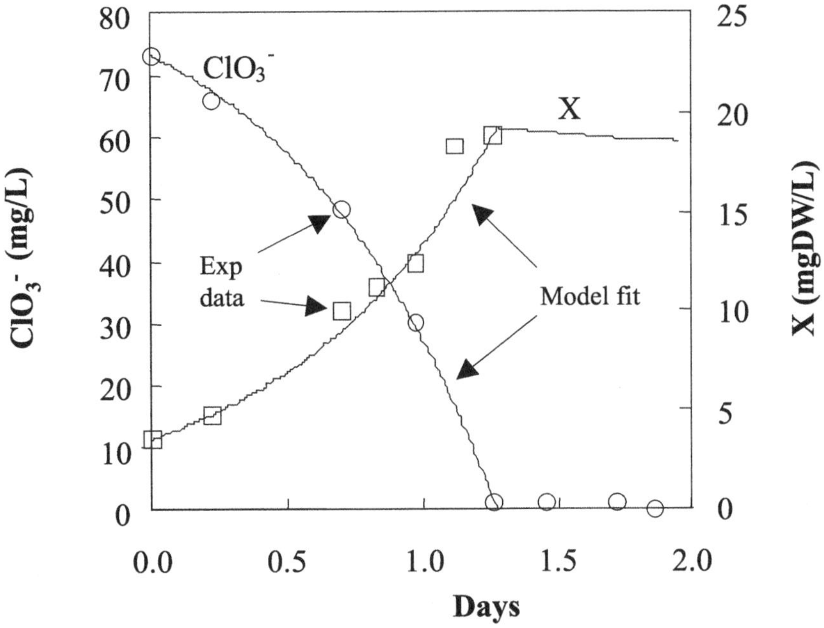

Figure 4.2 Typical growth experiment for chlorate (Model: q_{max} = 6.3 $gClO_3^-$/gDW-d, Y = 0.22gDW/$gClO_3^-$, and K = 1.2 $\mu gClO_3^-$/L)

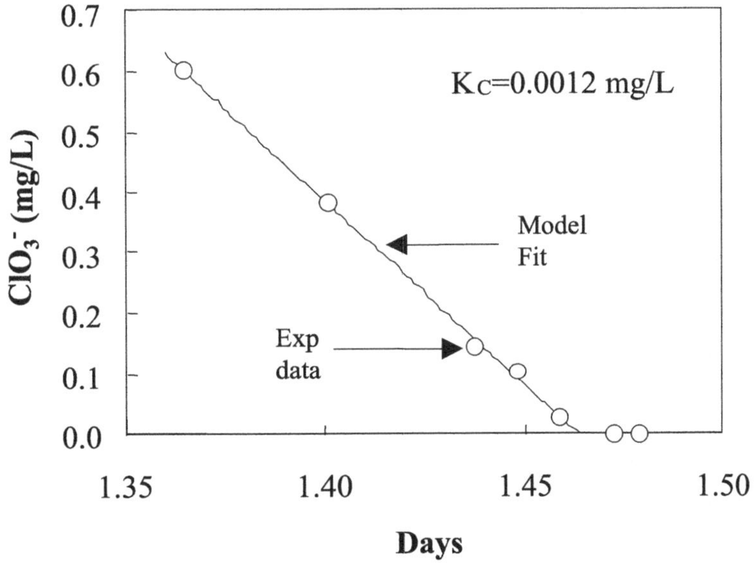

Figure 4.3 Typical K experiment for chlorate, non-growth conditions (Model: K = 1.2 µg ClO_3^-/L)

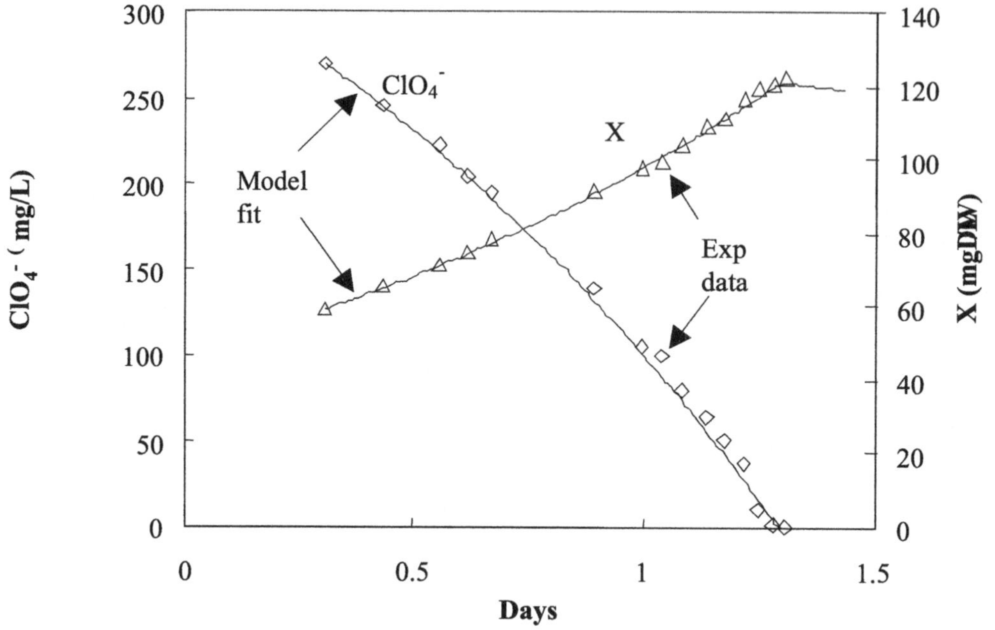

Figure 4.4 Typical growth experiment for perchlorate (Model: $q_{apparent}$ = 3.1 $gClO_4^-$/gDW-d, Y = 0.23gDW/$gClO_4^-$, and K = 0.15 $\mu gClO_3^-$/L)

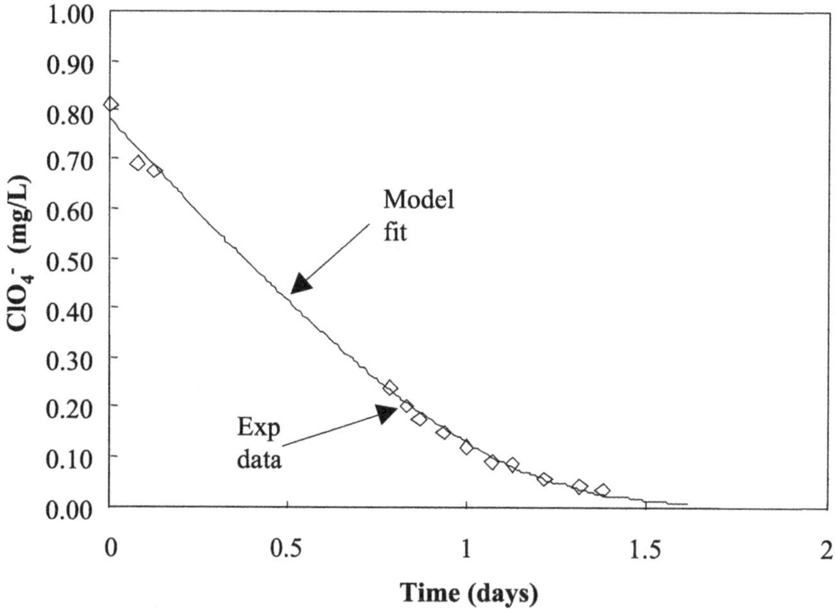

Figure 4.5 Typical K experiment for perchlorate, non-growth conditions (Model: K = 0.15 $\mu gClO_4^-$/L)

similar to the yield on chlorate, which is consistent with free energies per electron. Based on the McCarty method, as discussed previously for chlorate, the theoretical yield for perchlorate is 2.73 gDW/eq e^-. Thus, the determined yield for perchlorate is consistent with theoretical value. The yield for reduction of perchlorate to chlorate can be calculated by subtracting the per-electron chlorate yield multiplied by 6 electrons from the per-electron perchlorate yield multiplied by 8 electrons. The resulting yield is 0.055 mgDW/mgClO_4^-.

The q_{max} for perchlorate is 7.8 gClO_4^-/gDW-day. On a per-electron basis, the q_{max} is 0.63 e^- eq/mgDW-day, around twice that of chlorate. The apparent q_{max} for perchlorate, i.e., without considering inhibition, is 0.25 e^-eq/mgDW-day, or 3.1 mgClO_4^-/mgDW-day, which is less than for reduction of chlorate.

Other K values from the literature for PCRB are 33 and 12 mg/L (Logan et al., 2001), while the K for PC1, 0.15 mgClO_4^-/L, is around two orders of magnitude lower. PC1's low K for chlorate (1.2 μg/L) for PC1 explains the strong inhibition by chlorate on perchlorate reduction (Equation 4.7).

Nitrate

Typical growth curves and K tests for *Dechloromonas* sp. PC1 for nitrate are shown in Figure 4.6 and Figure 4.7. The true yields for nitrate is 0.88 gDW/gNO_3^--N. Expressed in terms of biomass per electron transferred to the acceptor, the yield for nitrate is 2.5 gDW/eq e^-. This is similar to chlorate and perchlorate, which is consistent with the similar amounts of free energy on a per-electron basis.

For autotrophic growth with hydrogen as an electron donor, nitrate as a nitrogen source, and no decay, the expected yield for nitrate is 2.63 gDW/eq e^-. Thus, the determined value for PC1 is consistent with theoretical values. The q_{max} value for nitrate is 4.0 gNO_3^--N/gDW-day. On the basis of electrons transferred to the acceptor, the q_{max} value is 1.4 e^-eq/mgDW-day. The nitrate reduction rate is more than three times of the rates of perchlorate and chlorate reduction on a per-electron basis.

The decay tests were carried with nitrate, and the results indicate a b value of 0.055 mgDW/mgDW-day, or 0.055 1/day. The maintenance requirements should the same for perchlorate, chlorate, and nitrate; therefore, this decay parameter value was used for all these acceptors.

Chlorate/Perchlorate Competitive Inhibitors

In order to determine the K_C for chlorate, a batch test was performed with initial equimolar amounts of perchlorate and chlorate, and another batch test with perchlorate in which accumulating chlorate was measured. The initial condition for the first test included 90-mg/L perchlorate and 75-mg/L chlorate. Perchlorate reduction was almost completely inhibited until chlorate was depleted (Figure 4.8). A low residual of chlorate remained until perchlorate was reduced. Using a q_{maxP} of 8.0 gClO_4^-/gDW-day and a q_{maxC} of 5.1 gClO_3^-/gDW-day, the best-fit K_C was 1.2 μg/L. The q_{maxC} was lower than the 6.0 gClO_3^-/gDW-day determined earlier in this section in order to fit the chlorate depletion. It is possible that the presence of perchlorate lowered the levels of enzyme expression with respect to growth with chlorate alone, although the rate for perchlorate was almost identical to that of the test with perchlorate alone. The test was highly sensitive to the K_C value: if K_C was too high, the perchlorate curve started to drop too gradually. If

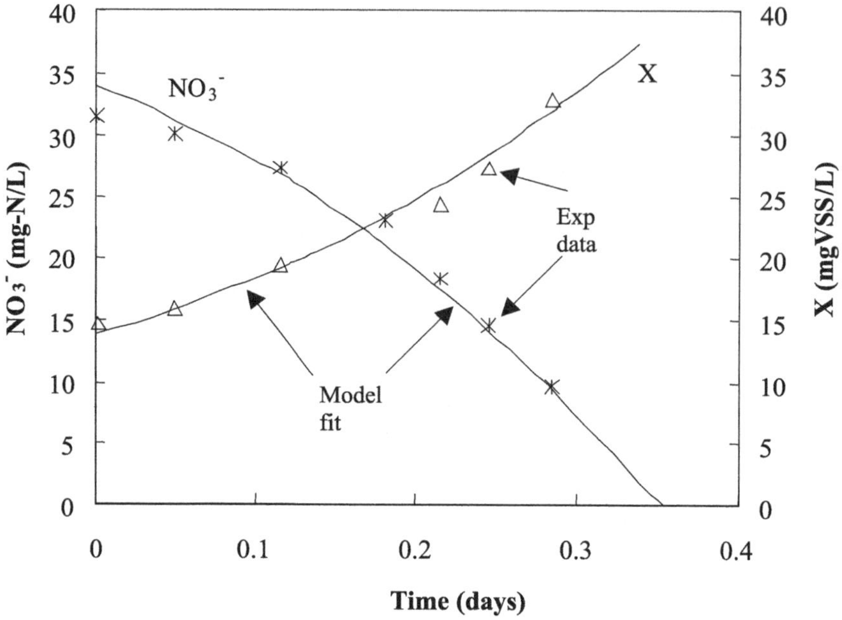

Figure 4.6 Typical growth experiment for nitrate (Model: q_{max} = 4.0 gNO_3^--N/gDW-d, Y = 0.88gDW/gNO_3^--N, and K = 0.30 $mgNO_3^-$-N/L)

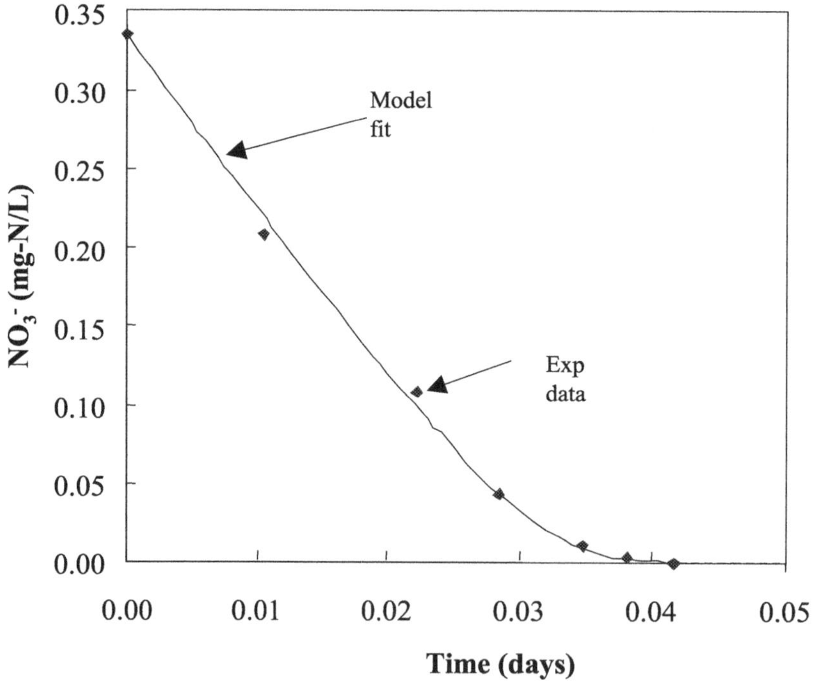

Figure 4.7 Typical K experiment for nitrate, non-growth conditions (Model: K = 0.30 $mgNO_3^-$-N/L)

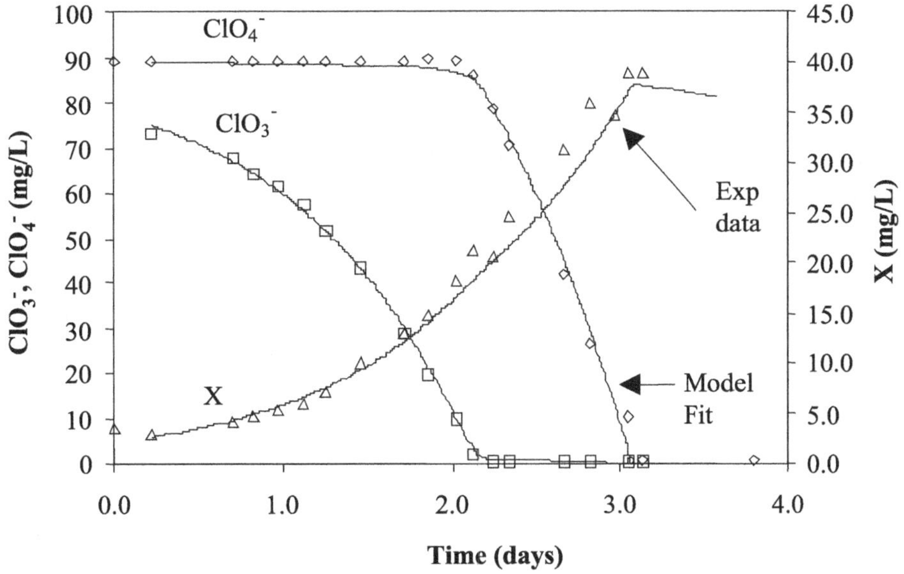

Figure 4.8 Experimental and modeled concurrent perchlorate and chlorate reduction (Model: q_{pmax} = 8.0 $gClO_4^-$/gDW-d, qc_{max} = 5.1 $gClO_3^-$/gDW-d, Y_P = 0.055 gDW/$gClO_4^-$, Y_C = 0.22 gDW/$gClO_3^-$, K_P = 0.15 $mgClO_3^-$/L, and K_C = 0.0012 mg/L)

K_C was too low, the residual amount of chlorate during perchlorate reduction was too low. An important result is that the competitive inhibition model was able to capture the shape of the chlorate and perchlorate curves shown in Figure 4.8 and Figure 4.9.

The determined K_C value was tested for a batch test with perchlorate alone, as shown in Figure 4.9. q_{max} values of 7.6 $gClO_4^-$/gDW-day and 4.5 $gClO_3^-$/gDW-day were used for perchlorate and chlorate, respectively. These values were slightly lower than the values determined in the above experiments. The K_C of 0.0012 mg/L allowed a good simulation of the accumulating chlorate, and the competitive inhibition model again accurately captured the shape of the perchlorate and biomass curves.

Summary and Analysis of Kinetic Parameters

The results of the various kinetic parameter experiments are summarized in Table 4.1.

Table 4.2 compares q_{max} and K values for perchlorate from the literature and from this study. The q_{max} values from the literature range from 1.68 to 7.5 $mgClO_4^-$/mgDW-day. Presumably, these are "apparent" values, as they were determined without considering competitive inhibition from chlorate. They are better compared to the apparent q_{max} for PC1, rather than the "true" value. The apparent q_{max} for PC1, 3.1 $mgClO_4^-$/mgDW-day, is near the low end of the range.

Growth Threshold Concentration, S_{min}

The kinetic parameters for perchlorate reduction can be used to determine the growth threshold concentration, S_{min}. S_{min} defines the lowest concentration of a substrate that provides a

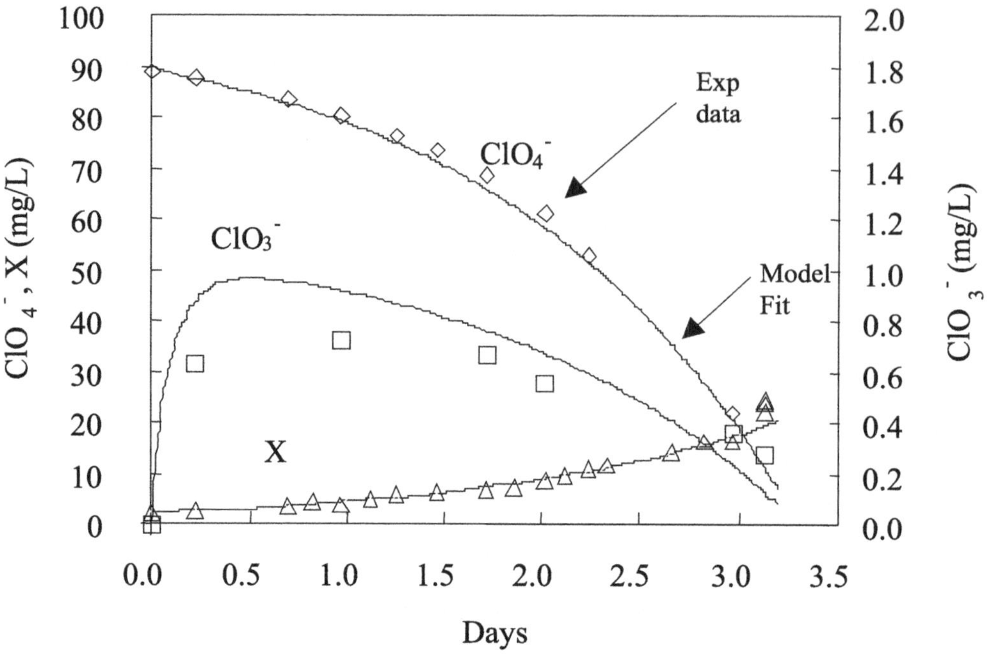

Figure 4.9 Batch growth with perchlorate, modeled with competitive inhibition (Model: q_{pmax} = 7.6 $gClO_4^-/gDW$-d, qc_{max} = 4.5 $gClO_3^-/gDW$-d, Y_P = 0.055 $gDW/gClO_4^-$, Y_C = 0.22 $gDW/gClO_3^-$, K_P = 0.15 $mgClO_3^-/L$, and K_C = 0.0012 mg/L)

Table 4.1
Kinetic parameters for *Dechloromonas* sp. PC1

	q_{max}		Y (for reduction to chloride)		
S	g acceptor/ gDW-day	(eq e^- H_2/ g X-day)	gDW/g acceptor	(gX/ eq e^- H_2)	K (mg/L)
ClO_4^-	7.8	0.63	0.05	2.9	0.15
ClO_3^-	6.0	0.31	0.22	2.8	0.0012
NO_3^-	4.0	1.4	0.88	2.5	0.030

NOTES:
(1) "eq e^- H_2" = equivalent of electrons from hydrogen
(2) 1 eq e^- H_2 = 1 g H_2
(3) b = 0.1 1/day
(4) $\Delta G_o'$ = Gibb's free energy at standard conditions and pH = 7

Table 4.2
Apparent kinetic parameters for PCRB

Isolate	Acceptor	q_{max} mg acceptor $mgDW^{-1}day^{-1}$	K mg/L	Reference
Vibrio dechloratans	Perchlorate	1.68	—	Korenkov et al. (1976)
Wolinella succcinogenes HAP-1	Perchlorate	2.57	—	Wallace et al. (1996, 1998)
GR-1	Perchlorate	5.65	—	Rikken, Kroon, and van Ginkel (1996)
GR-1	Chlorate	7.48	—	Rikken, Kroon, and van Ginkel (1996)
KJ	Perchlorate	24	33	Logan et al. (2002)
PDX	Perchlorate	7.5	12	Logan et al. (2002)
PC1	Perchlorate	3.1 (7.8)	0.15	This study
PC1	Chlorate	4.1	0.0015	This study

The value in parenthesis is the rate considering competitive inhibition from chlorate

non-negative growth rate (Rittmann and McCarty, 2001). It can be applied to biofilms or suspended biomass. Consider the biomass balance for batch growth:

$$\frac{dX}{dt} = q_{\max}\frac{S}{S+K}YX - bX \tag{4.10}$$

When S is small with respect to K, it can render the positive term on the right side of the equation smaller than the negative term, providing a net decay in biomass for any value of X. S_{min} is the minimum concentration that can support steady-state biomass for a continuous suspended or biofilm system, and is derived by setting the growth rate (*dX/dT*) to zero, resulting in:

$$S_{\min} = \frac{Kb}{Yq_{\max} - b} \tag{4.11}$$

For perchlorate as a limiting acceptor, competitive inhibition may affect the S_{min}. Neglecting biomass produced by changes in chlorate concentrations, which is insignificant compared that from perchlorate reduction to chloride, the biomass equation can be written as follows:

$$\frac{dX}{dt} = q_{\max P}\frac{S_P}{S_P + K_P\left(1 + \frac{S_C}{K_C}\right)}YX = bX \tag{4.12}$$

The S_{min} is determined by setting this equation to zero and isolating S_P. For PC1, at low perchlorate concentrations, the term in parenthesis in the denominator tends towards 1, and the biomass equation without competitive inhibition can be used. The resulting S_{min} expression is:

$$S_{\min P} = \frac{K_P b}{Y q_{\max P} - b} \qquad (4\text{-}13)$$

where

q_{maxP} is expressed in mg/mgDW-day
Y is the yield for complete perchlorate reduction (i.e., to chloride)

If b is 0.055 mgX/mgX-day, K_P is 0.15 mg/L, Y is 0.05 mgDW/mgClO$_4^-$, and q_{Pmax} is 7.8 g ClO$_4^-$/gDW-day, the S_{min} is 5 μgClO$_4^-$/L. For comparison, the S_{min} for chlorate and nitrate are 6.5×10^{-2} μg ClO$_3^-$/L and 4.3×10^{-4} μg NO$_3^-$/L, respectively. For a biofilm process, the b is replaced by b′, which includes detachment in addition to maintenance. Typical detachment values range from 0.01 to 0.1 mgX/mgX-day (Rittmann and McCarty, 2001), and previous studies with the MBfR gave very low detachment values, around 0.01 mgX/mgX-day (Lee and Rittmann, 2001; Lee and Rittmann, 2002). For the MBfR, the total b′ would be 0.065 mgX/mgX-day, providing an S_{min} for perchlorate is of 7 μg/L. For a reactor with a detachment rate of 0.1 mgX/mgX-day, b′ would be 0.155 mgX/mgX-day, and the resulting S_{min} would be 14 μg/L.

Conclusions

PC1 is an autotrophic, hydrogen-oxidizing PCRB. It is a Gram-negative rod and can grow with nitrate, nitrite, and oxygen, but not with bromate, arsenate, selenate, selenite, or sulfate. It belongs to the genus *Dechloromonas*, which includes many PCRB.

Kinetic parameters were determined for PC1 under autotrophic, hydrogen-oxidizing conditions. The K for perchlorate was 0.15 mg/L, two orders of magnitude lower that found for other bacteria, suggesting that this isolate can remove perchlorate to lower levels than other PCRB with similar q and b values. Chlorate and perchlorate mutually inhibit each other, and a competitive inhibition model can describe this behavior. The inhibition can have a significant effect on the perchlorate reduction rates, even when the chlorate concentration is very low. The effect should be constant over a wide range of perchlorate concentrations; therefore, an "apparent" q_{max} determined without competitive inhibition can be used to determine the actual q_{max}. When the perchlorate concentration is in the μg/L range, however, competitive inhibition is expected to have a minor effect on reduction rates. Therefore, using the apparent q_{max} at such concentrations could significantly underestimate reduction rates.

The S_{min} for PC1 for perchlorate under autotrophic conditions with hydrogen is at least 5 μg/L for the MBfR. Although this is nearly equal to the 4 μg/L standard, this is an extreme condition, and it would be difficult and inefficient to operate a reactor at S_{min}. Achieving perchlorate concentrations less than 4 μg/L may require that perchlorate is a secondary substrate, a concept explained in the next chapter.

CHAPTER 5
MICROBIAL ECOLOGY INVESTIGATIONS

Understanding how and why perchlorate-reducing bacteria become a significant fraction of the biomass is key for reliable reactor operation, particularly when the influent perchlorate concentrations are very low. It is also important to know how much biomass is capable of reducing perchlorate in order to predict process performance. Integral to both of these key points are the bacteria themselves. Consequently, the project team has also endeavored to identify the specific organisms responsible for the perchlorate reduction.

Preliminary investigations presented in the previous chapters suggest that, when an MBfR is fed nitrate alone, the microbial community consists of denitrifying bacteria (DNB) and a small proportion of perchlorate reducing bacteria (PCRB). When perchlorate is added, the proportion of PCRB increases. For example, in the mechanisms experiments, an MBfR was supplied with 5-mg-N/L nitrate and allowed to reach steady state. On adding 1 mg/L perchlorate to the feed, removals were only 30 percent of the influent perchlorate. However, the removal increased to 99 percent over the next two weeks, suggesting enrichment for PCRB. Another example is an experiment performed to test a defined medium that appeared to be favorable for DNB, but unfavorable for PCRB. Perchlorate reduction was nearly lost over several months with the defined medium, but returned to 99 percent over a two-week period when the original tap-water medium was restored. Despite the drastic effect on perchlorate, the defined medium had no appreciable effect on nitrate removals.

Although these data suggest that perchlorate enriches for PCRB, it is not clear how perchlorate brings about the enrichment. If perchlorate were the sole acceptor used by PCRB, the number of PCRB would be governed by the amount of perchlorate reduced and the biomass yield for perchlorate reduction. Since the amount of perchlorate in drinking water is typically very low relative to nitrate, the amount of PCRB would be expected to be insignificant compared to DNB. This is illustrated in Figure 5.1, which presents the percentages of biomass based on combined theoretical yields on 8 mg/L oxygen, 5 mg-N/L nitrate, and 10,000, 1,000, or 100 µg/L perchlorate. The theoretical yields, based on the McCarty thermodynamic method (Rittmann and McCarty, 2001) are of 0.28 mg DW/mg ClO_4^-, 1.1 mg DW/mg NO_3^--N, and 0.10 mg DW/mg O_2. With 10,000 µg/L perchlorate, the biomass grown solely from perchlorate reduction is around 22 percent of the total. With 1,000 µg/L perchlorate, the fraction drops to 3 percent. With 100 µg/L, which is more representative of perchlorate-contaminated drinking water, the fraction is only 0.3% percent. Of course, with no influent perchlorate, the number of PCRB should be zero if perchlorate were the only acceptor respired by the PCRB.

Two factors may account for higher numbers of PCRB than predicted by the relative yield from perchlorate reduction alone. Both factors are related to the fact that many PCRB are denitrifiers, while all identified PCRB can use oxygen. One potential explanation for the prevalence of a PCRB under denitrifying or oxygen-depleted conditions is that, a given PCRB may be more efficient at denitrification and oxygen reduction than non-perchlorate-reducing bacteria. Another potential explanation is that when oxygen and nitrate concentrations are driven to very low levels inside a reactor, the ability to use perchlorate may give PCRB an advantage that helps them out-compete the other DNB for nitrate and oxygen. The latter strategy, which seems more likely, is based on perchlorate being utilized as secondary acceptor. The same strategy

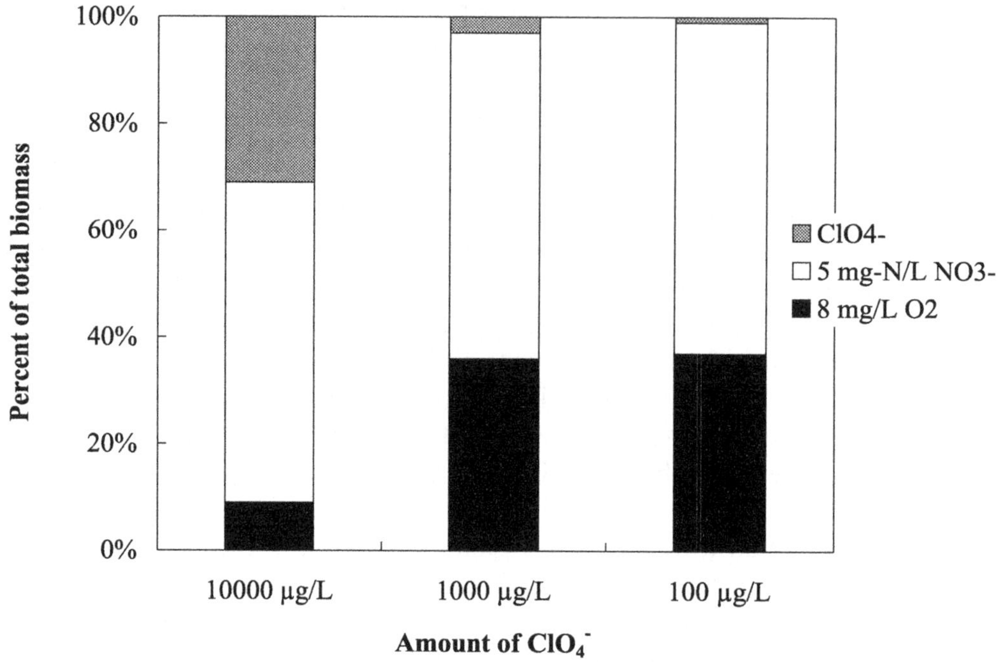

Figure 5.1 Percentages of hydrogen-oxidizing biomass generated by 8-mg/L oxygen, 5-mg-N/L nitrate, and various amounts of perchlorate

explains why PCRB can be present when no perchlorate is available and why the perchlorate concentration can be driven below S_{min} for steady-state operation of the MBfR.

STRATEGY FOR MICROBIAL ECOLOGY EXPERIMENTS

To evaluate the microbial ecology of the MBfR under controlled conditions, five bench-scale MBfRs were prepared. Four were used to study perchlorate reduction in a denitrifying community (i.e., with nitrate as a primary acceptor), and the fifth was used to study perchlorate reduction with oxygen as a primary acceptor. All five reactors were identical and run concurrently. Precautions were taken to ensure that conditions (e.g., medium flow-rate, recirculation rate, pH, temperature) were the same among all reactors so that changes in ecology resulted from the different electron acceptors and not from other environmental factors.

MATERIALS AND METHODS

Reactor Configuration

The reactor configuration was similar to that used in the preliminary experiments, as described in Chapter 4, except at a smaller scale and around one-tenth of the flow rate. The typical configuration of the MBfR used in these microbial ecology experiments is shown in Figure 5.2. Table 5.1 summarizes the physical characteristics of the reactors.

Each of the five ecology reactors consisted of two glass tubes connected with Norprene tubing and plastic "tee" fittings. One glass tube contained a main bundle of 32 hollow-fiber membranes, each with a 25-cm active length. A second manifold was constructed with a single

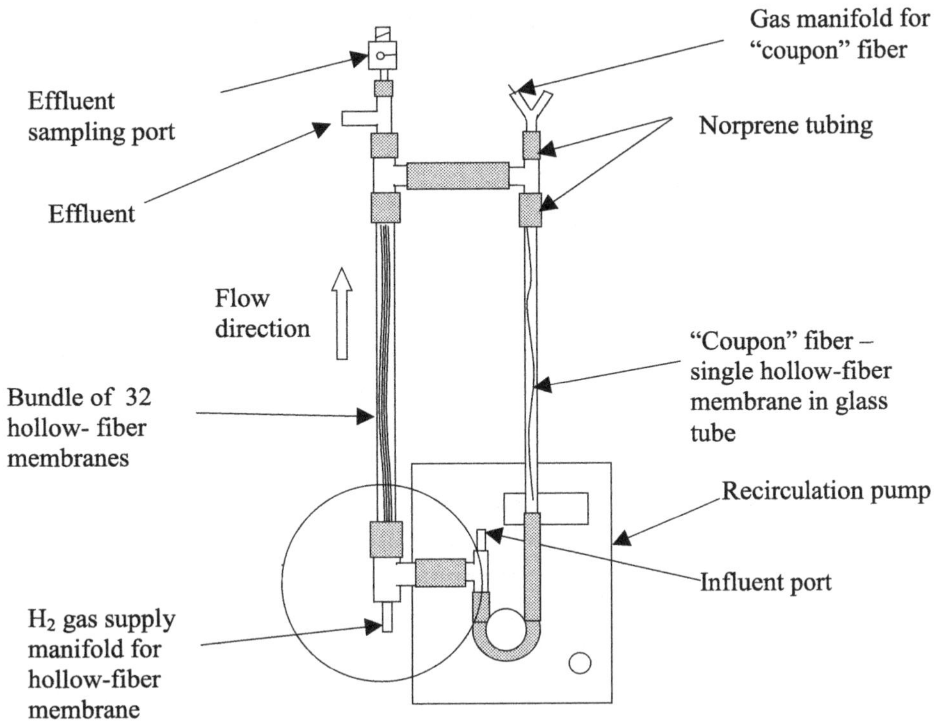

Figure 5.2 Typical bench-scale MBfR configuration

Table 5.1
MBfR reactor characteristics for the ecology experiments

	Units	Main tube	Coupon tube	Reactor total
Tube inside diameter	cm	0.6	0.5	—
No. of hollow fibers		32	1	33
Cross-sectional area fibers	cm^2	0.019704	0.000616	0.02032
Feed rate	ml/min	—	—	1
Recirculation rate	ml/min	—	—	150
Net cross sectional area	cm^2	0.26	0.20	—
Fiber surface area	cm^2	70.37	2.20	72.57
Liquid velocity	cm/min	570.3	766.3	—
Average detention time	min	—	—	23.9

fiber. The purpose of the fiber, which was 25-cm long, was to act as a "coupon" that could be sampled without disturbing the main bundle of fibers and without causing a significant change in the reactor surface area. Removing three 2.5-cm samplings from the coupon would decrease the reactor's total fiber surface area by less than 1%. The reactor's high recirculation rate, 150 mL/min, promoted completely mixed conditions inside the reactor. It also ensured that all points of the bulk liquid inside the reactor were experiencing concentrations similar to the effluent

Table 5.2
Electron acceptors in bench-scale reactors

Reactor	O_2 (mg/L)	NO_3^- (mg-N/L)	ClO_4^- (μg/L)
R1	6	5	0
R2	6	5	100
R3	6	5	1,000
R4	6	5	10,000
R5	8	0	1,000

Table 5.3
Reactor medium

Reactor medium (per liter)	
0.128 g	KH_2PO_4
0.434 g	Na_2HPO_4
0.2 g	$MgSO_4{\cdot}7H_2O$
0.5 mL	NO_3^- stock (10 gN/L)*
1 mg	$CaCl_2{\cdot}2H_2O$
1 mg	$FeSO_4{\cdot}7H_2O$
1 mL	Trace mineral solution

* For oxygen reactor, no nitrate is used and 0.2 mg/L $(NH_4)_2SO_4$ is used as a nitrogen source instead.

concentration. For example, a reactor with influent flow rate of 1 mL/min, a 150-mL/min recirculation rate, a 5-mg/L influent concentration, and a 0.05-mg/L effluent concentration has a 0.08-mg/L (1 mL/min × 5 mg/L / 150 mL/min + (150-1) mL/min × 0.05 mg/L / 150 ml/min) concentration at the point where the medium is input into the reactor. All other points in the reactor's bulk liquid would have concentrations between 0.05 and 0.08 mg/L.

Another advantage of a high recirculation rate is the high hydraulic shear on the fibers, which encourages the development of a dense biofilm (Chang et al., 1991). It also minimizes the accumulation of excessive biomass that can clog the reactor.

Reactor Media

Table 5.2 summarizes the electron acceptors supplied in the influent of each of the five MBfRs. The medium used for the four nitrate reactors is described in Table 5.3, and the trace mineral solution is described in Table 5.4. The nitrate medium was prepared in a 20-L plastic drum (Nalgene, Rochester, NY) and filter sterilized into another sterile 20-L drum using a capsule filter (Pall SuporCap 100, Pall Corporation, Ann Arbor, MI). Each capsule filter contains an 80-μm pre-filter followed by a 20-μm filter. The first drum was pressurized with N_2 gas to force

Table 5.4
Trace mineral solution

Trace mineral solution (mg/L)	
$ZnSO_4 \cdot 7H_2O$	100
$MnCl_2 \cdot 4H_2O$	30
H_3BO_3	300
$CoCl_2 \cdot 6H_2O$	200
$CuCl_2 \cdot 2H_2O$	10
$NiCl_2 \cdot 6H_2O$	10
$Na2MoO_4 \cdot 2H_2O$	30
Na_2SeO_3	30

Table 5.5
Summary of reactor operation

		Acceptors present in influent			Biofilm
Day	Feed (mL/min)	R1	R2-R4	R5	sampling
0	0	$O_2 + NO_3^-$	$O_2 + NO_3^-$	O_2	
1	1.0	$O_2 + NO_3^-$	$O_2 + NO_3^-$	O_2	
20	1.0	$O_2 + NO_3^-$	$O_2 + NO_3^-$	O_2	√
20	1.0	$O_2 + NO_3^-$	$O_2 + NO_3^- + ClO_4$	$O_2 + ClO_4^-$	
36	1.0	$O_2 + NO_3^-$	$O_2 + NO_3^- + ClO_4$	$O_2 + ClO_4^-$	√
40	0.85/1.0	$O_2 + NO_3^-$	$O_2 + NO_3^- + ClO_4$	$O_2 + ClO_4^-$	
71	0.85/1.0	$O_2 + NO_3^-$	$O_2 + NO_3^- + ClO_4$	$O_2 + ClO_4^-$	√

the medium through the filter. The media used for the oxygen reactor was identical to the nitrate media, except that it had no nitrate, but NH_4^+ as a nitrogen source. Both media had a phosphate buffer designed to provide a final pH of around 7.4.

Reactor Operation

The five reactors were peroxide sterilized prior to inoculation with a sample of the pilot-scale MBfR discussed later in Chapter 6. Once the reactors were seeded, influent media was pumped to each reactor at a rate of 1 mL/min. The effluent nitrate for all nitrate reactors reached steady state after around 3-5 days. After 20 days, a fiber coupon section was harvested, and perchlorate was added to the feed to reactors R2 through R5. After operating with perchlorate for 16 days, a second fiber coupon section was harvested. Four days after harvesting the coupon, the influent flow rate for reactors R1-R4 was decreased to 0.85 mL/min, while R5 was maintained at 1 mL/min. At 51 days after starting the perchlorate feed, another coupon was sampled. The reactor operation is summarized in Table 5.5.

Table 5.6
Culturability of bacteria, determined as a percentage of culturable bacteria in comparison with total cell counts (Amann, Ludwig, and Schleifer, 1995)

Habitat	Culturability (%)*
Seawater	0.001-0.1
Freshwater	0.25
Mesotrophic lake	0.1-1
Unpolluted estuarine waters	0.1-3
Activated sludge	1-15
Sediments	0.25
Soil	0.3

* Culturable bacteria are measured as colony forming units

Analytical Methods

The analytical methods for perchlorate, nitrate, liquid-phase hydrogen, pH, COD, and dissolved oxygen are the same as those presented previously in Chapter 4 except that a Dionex AS-11 column with an AG-11 precolumn and a 200 μL injection loop was used for perchlorate analysis. The perchlorate detection level with this method was 15 μg/L.

Molecular Methods

Traditional methods of assaying biomass, such as microscopic enumeration and most probable number determinations, have serious limitations for studying microbial ecology. Since PCRB and DNB can have the same morphologies, it is not possible to quantify each type by microscopic enumeration. Most probable number (MPN) determinations (Koch, 1994) can give an approximate value of relative numbers of PCRB and DNB, but this method is limited by the culturability of the PCRB and DNB in the biofilm. It is well known that only a small fraction of bacteria are culturable (Amann, Ludwig, and Schleifer, 1995), as illustrated in Table 5.6. Therefore, the results of MPN tests can be biased towards the group with more culturable bacteria, as opposed to total bacteria.

Molecular techniques were developed to overcome the bias of the traditional culture-based methods (Ward, Weller, and Bateson, 1990; Spring et al., 1992; Muyzer, Dewaal, and Uitterlinden, 1993; Amann, Ludwig, and Schleifer, 1995). These techniques can quantify the amount of a gene or gene product that is found in a microbial community without the need to culture each organism. Two powerful molecular tools were used to study microbial communities: Denaturing Gradient Gel Electrophoresis (DGGE) and Fluorescent *In-situ* Hybridization (FISH), which are described below.

Based on the strengths and limitations of DGGE and FISH, a strategy that combines both techniques was devised. The strategy uses DGGE to compare the ecology of reactors with equal amounts of oxygen and nitrate, but with different amounts of perchlorate, and it uses FISH to quantify the abundance of the species of interest. It was expected that DGGE bands corresponding to PCRB would increase in intensity in the reactors having prolonged exposure to higher perchlorate inputs. Extraction and sequencing the DNA from these bands would allow for the design FISH probes to quantify the species of interest *in-situ*. Probes could also be designed based on PCRB isolated from the inoculum.

DNA Extraction

DNA was extracted from the biofilm sample and from bacterial isolates using the UltraClean Soil DNA Isolation Kit and the UltraClean Microbial DNA Isolation Kit (Mo Bio laboratories, Carlsbad, CA), respectively. The kits were used according to the manufacturer's instructions.

DGGE Analysis

DGGE is one of several RNA- and DNA-based nucleic acid fingerprinting techniques used to characterize a mixed microbial community. The technique is based on the extraction and amplification via polymerase chain reaction (PCR) of specific regions of nucleic acid, with a length limited to around 500 base pairs (Muyzer, Dewaal, and Uitterlinden, 1993; El Fantroussi et al., 1999; Koizumi et al., 2002). As opposed to agarose gel electrophoresis, which separates DNA fragments based on size, DGGE separates fragments of equal size based on their difference in sequence (Myers et al., 1985; Wilderer et al., 2002). In DGGE, bacteria in a mixed culture are lysed to release their DNA. Primers are used to amplify the DNA regions of interest, and the amplified DNA is loaded onto a polyacrylamide gel with a linearly increasing gradient of chemical denaturant, usually formamide plus urea. The primers used in the PCR have an added GC-rich region. The GC bond is stronger than that of AT, and a series of GCs is highly resistant to denaturing, providing a "clamp" that holds the DNA strands together. When variable portions of the DNA strand reaches a point in the gel where it denatures, the GC clamp prevents the two strands from separating. The open portion of the DNA, held together by the clamp, acts like a parachute and stops the DNA from further migration. All the copies of DNA with the same sequence stop at the same position, so each bacterium should have a distinct band location. The bands can be stained with a DNA binding agent, such as ethidium bromide, and viewed on an ultra violet (UV) transilluminator.

DNA/RNA fingerprinting methods, such as DGGE, are ideal to assess shifts in a microbial population during changes in environmental conditions (Muyzer, Dewaal, and Uitterlinden, 1993; Muyzer, 1999). However, due to the biases associated with DNA amplification via PCR (Suzuki and Giovannoni, 1996), they are not considered reliable for quantification of or determining the absolute abundance of different members of a microbial community. A powerful advantage of DGGE over other DNA/RNA fingerprinting methods, such as Terminal Restriction Fragment Length Polymorphism (T-RFLP) (Liu et al., 1997; Moeseneder et al., 1999), is that bands of interest may be extracted and sequenced for phylogenetic identification.

The 16S rDNA genes from extracted biofilm samples and bacteria isolates were amplified by Polymerase Chain Reaction (PCR) using PCR Master (Roche Applied Science, Indianapolis, IN) for DGGE analysis according to the manufacturer's instructions. The universal primer combination was GM5F-GC (forward primer) and DS907R (reverse primer). The forward primer GM5F-GC contained a 40-base GC clamp. PCR was performed using and automated thermal cycler (PCR Express, Hybaid, Ulm, Germany) by a "touchdown" PCR method (Giovannoni, 1991). The amplified DNA fragment, around 550-base pairs (bp), was analyzed by DGGE.

Table 5.7
Nucleic acid primers used for PCR amplification of 16S rDNA

Primer	Sequence	Reference
GM5F-GC	5'-CGCCCGCCGCGCGCGGCGGGCGGGGCGG GGGCACGGGGGGCCTACGGGAGGCAGCAG-3'	Teske et al. (1996)
DS907R	5'-CCCCGTCAATTCCTTTGAGTTT-3'	Teske et al. (1996)
U27F	5'-AGAGTTTGATCCTGGCTCAG-3	Shinoda, Kato, and Morita (2000)
U1112F	5'-GTCCCGCAACGAGCGCAAC-3'	Shinoda, Kato, and Morita (2000)
U1525R	5'-AAAGGAGGTGATCCAGCC-3'	Shinoda, Kato, and Morita (2000)
U805F	5'-GATTAGATACCCTGGTAGTC-3'	Shinoda, Kato, and Morita (2000)
U518R	5'-GTATTACCGCGGCTGCTGG-3'	Shinoda, Kato, and Morita (2000)
U1093R	5'-TTGCGCTCGTTGCGGGACT-3'	Shinoda, Kato, and Morita (2000)

DGGE was performed using a BioRad Dcode system (Bio Rad, Hercules, CA). an acrylamide gel (6% [w/v]) was prepared and run in a 0.5× TAE buffer (20mM Tris-acetate [pH7.4], 10mM Acetate, 0.5mM Na_2EDTA). The denaturing gradient ranged from 30 to 60% denaturant (100% denaturant corresponds to 7M urea and 40% [v/v] formamide). Electrophoresis was run at a constant voltage of 75V at 60°C for 16 hours. The gel was stained with SYBR Green I (Molecular Probes, Eugene, OR) at 10000× dilution in DI water and viewed the stained gel on a UV trans illuminator. The gel image was captured using a CCD camera system (Cohu Electronics model 4912-2010, San Diego, CA).

DNA Sequencing of DGGE Bands and Bacterial Isolates

DGGE bands were excised with a sterile pipette tip and transferred to sterile 0.2-ml microcentrifuge tubes. After rinsing twice with 50 μl of sterile, deionized water, the excised gel were used for re-amplification by PCR. A primer combination of GM5F-GC and DS907R were used under the same conditions as described above for DGGE analysis. The PCR products were purified using the UltraClean PCR Clean-up Kit (Mo Bio laboratories, Carlsbad, CA) and carried out with cycle sequencing using an ABI 3100 Prism BigDye terminator sequencing standard kit (Applied Biosystems, Foster City, CA) with the primer GM5F-GC or DS907R, according to the manufacturer's instructions.

The partial sequence of 16S rDNA of isolated bacteria was amplified by PCR with universal primer combination U27F and U1525R (Table 5.7) and purified the amplified DNA fragment using UltraClean PCR Clean-up Kit (Mo Bio laboratories, Carlsbad, CA) and supplied for cycle sequencing. The cycle sequencing was performed by the same procedure as described above with seven universal primers: U27F, GM5F, U805F, U518R, U1093R, U1112F, and U1525R (Table 5.7). The DNA sequence was determined using an ABI 3100 DNA sequencer after purification of cycle sequence-product by a conventional ethanol-precipitation method.

The DNA sequence of a DGGE band or isolated bacterium was compared with the FASTA DNA database (Mackey, Haystead, and Pearson, 2002; Miyazaki et al., 2003). Sequence alignment and phylogenetic analysis was performed using the ClustalX software (Thompson et al., 1997).

FISH

FISH is a molecular technique that allows *in-situ* identification of bacteria without the need to extract and amplify the genes or gene products. FISH targets the 16S rRNA, which is part of the ribosome, a protein-synthesing organelle (Madigan, Martinko, and Parker, 1997). Bacteria cells contain thousands of copies of the 16S rRNA, with the exact number depending on the growth conditions (Madigan, Martinko, and Parker, 1997). For example, cells at exponential growth usually have more 16S rRNA than cells at the stationary growth phase (Binder and Liu, 1998; Binnerup et al., 2001). The 16S rRNA is useful for distinguishing among bacterial groups, because it has highly conserved regions necessary for ribosome function and variable regions that vary only slightly in closely related organisms and greatly for distantly related organisms (Stahl and Amann, 1991). As a result, oligonucleotide probes can be designed to be complementary to unique regions of the 16S rDNA of the organism or group of organisms of interest (Manz et al., 1992; Amann, Ludwig, and Schleifer, 1995). Probes can be designed to differentiate bacteria at different levels of specificity; for example, by species, genus, kingdom, or for all life. The 16S rRNA is present in sufficient quantity to enumerate even small populations.

A limitation of FISH for differentiating and counting PCRB is that the PCRB are not phylogenetically coherent (Coates et al., 1999). Thus, different strains of PCRB are not necessarily closely related and, in fact, may be very closely related to bacteria that do not reduce perchlorate.

Two oligonucleotide probes were designed, one for *Dechloromonas* CH1, an isolate obtained from the MBfRs, and the second for PB9, an uncultured species that was dominant in our DGGE experiments and whose sequence was determined from its DGGE band. The probe sequence was determined by using the PRIMROSE software (Ashelford, Weightman, and Fry, 2002), which is a probe-design tool, with the 16S rRNA sequence of *Dechloromonas* species enrolled in Ribosomal Database Project II (Cole et al., 2003).

The probe sequences were as follows:

CH1: 5'-GCCCATGCGATTTCTTCC-3'
PB9: 5'-ACCTATGCGATTTCTTCCC-3'

The optimal formamide concentration for hybridization was determined by using extracted DNAs from CH1 and the mixed-culture biofilm, testing them with 20 to 55 percent formamide concentrations. When the formamide concentration was 40 percent, PB9 did not hybridize with CH1, and also the ratio of number of CH1-hybridized bacteria toward others in the extracted DNA from biofilm declined remarkably. Therefore, 35% formamide was used for simultaneous FISH with both probes.

The biofilm was fixed with 1 mL of 4 percent formaldehyde solution in 1.5-ml microcentrifuge tube for 18 hours, the formaldehyde solution was discarded, and 1-mL of 1:1 mix of PBS/ethanol solution was added. After this step, the specimen was preserved at –20°C. The biofilm was dispersed by sonication, with periodic cooling on ice, for approximately 5 minutes (Aquasonic sonicator model 250 HT, VWR Scientific, West Chester, PA), and mechanical disruption by passing the specimen through a 200 µL pipette tip. Then the sample was spotted onto a 20-µL aliquot on a gelatin-coated microscope slide and dehydrated in a 50, 80, and 96 percent ethanol series. After air-drying the slide, 10-µL of hybridization buffer was applied (900 mM NaCl, 20mM Tris/HCl, 35 percent formamide, 0.01 percent SDS), including 2ng/µL of each probe:

(1) PB9 labeled with Cy3 (sulfoindocyanate) at 5' end,
(2) CH1 labeled with FITC (fluorescein isothiocyanate) at 5' end.

The hybridization step was performed at 46°C for 2 hours, according to the procedure described by Manz et al. (1992) and Pernthaler (1997). After hybridization, the slide was immediately poured into 50-mL of pre-warmed washing buffer (52mM NaCl, 20mM Tris/HCl, 5mM EDTA, 0.01 percent SDS) and incubated at 48°C for 15 minutes. The slide was washed with distilled water twice and air-dried. For the counter-stain, 20 µL of 1 percent DAPI (4,6-diamidino-2-phenylindole) solution was spotted and the slide was incubated for 3 minutes. Then the slide was washed with 80 percent ethanol and distilled pure water to remove non-specific staining. The specimen was mounted in Citifluor and observed with an Axiophot epifluorescence microscope (Carl Zeiss, Thornwood, NY). Images were captured with an Axiocam CCD (Carl Zeiss, Thornwood, NY) using the provided software. The stained colonies for each probe were counted for several different frames, and two different biofilm locations were counted. Colonies that were stained with the PB9 or the CH1 probe were compared to the total number of bacteria strained with the DAPI stain.

Isolation of Pure Cultures

Attempts were made to isolate pure cultures of perchlorate-reducing bacteria from the inoculum and from the reactor effluent. Bacteria were enriched from the 160-mL serum bottles filled with 25-mL growth medium with perchlorate, as described previously, and a 95:5 v/v mix of hydrogen and carbon dioxide. Then, the enrichment was plated aerobically in R2A agar (Difco, Detroit, MI). Biofilm also was plated from the reactor effluent and from coupons. The biofilm collected from reactor effluent or from the coupon was disrupted mechanically by pipetting with a sterile, 200-µL pipette. All suspensions were diluted between 10^3 and 10^7 and plated by adding 100-µL to a plate and spreading uniformly. The plates were incubated at 22°C. Colonies were picked from the plates and re-spread to purity. Then, their metabolism was tested by growing them in Balch tubes with 5-mL of media and using oxygen, nitrate, or chlorate as the acceptor. The bottle was vacuum-degassed to remove air, and 95:5 v/v mix of hydrogen and carbon dioxide was supplied to the headspace. For growth with oxygen, 5-mL of pure oxygen was added to the headspace. Those bacteria growing on chlorate were also tested on perchlorate. Autotrophic and heterotrophic tubes were used.

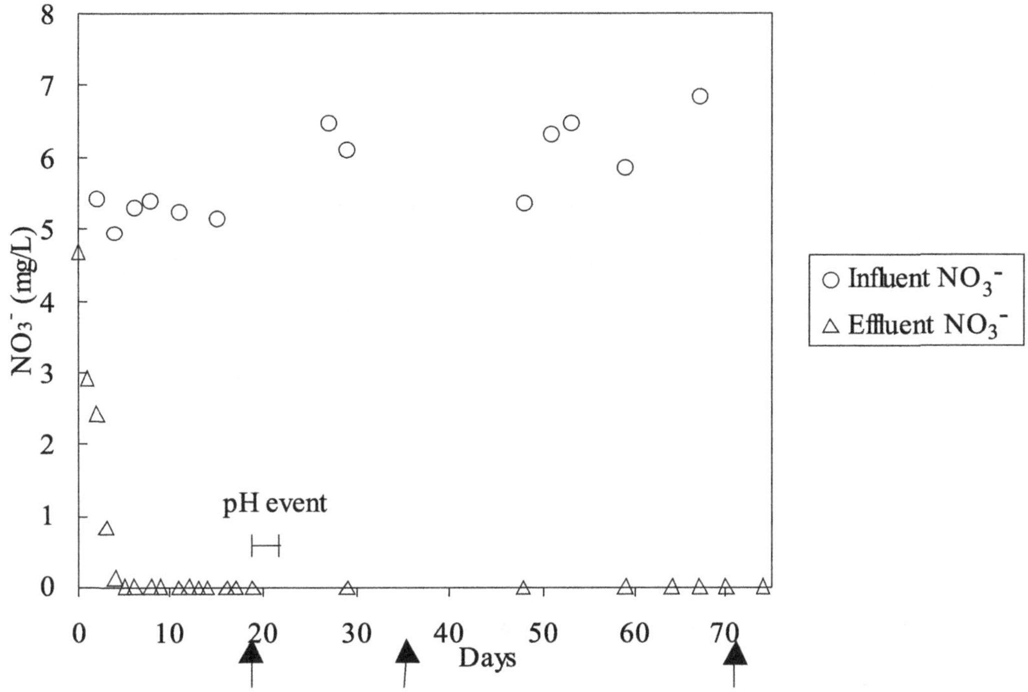

Figure 5.3 Reactor R1 influent and effluent nitrate concentrations (arrows indicate coupon-sampling dates)

RESULTS AND DISCUSSION

Reactor Operation

After the initial seeding, reactors R1 through R4 were supplied with a feed medium having 6-mg/L oxygen and 5-mg-N/L nitrate, while the feed to reactor R5 contained 8-mg/L oxygen. The oxygen and nitrate concentrations in the effluent dropped to non-detect and around 30-μgN/L, respectively, within five days. Figure 5.3 shows the influent and effluent concentrations for R1. For reactors R1-R5, the typical effluent pH was 7.4, the typical hydrogen concentration was 0.3-0.5 mg/L, and the effluent COD was 10 mg/L (unfiltered) and 6 mg/L (filtered). The influent COD was less than 1 mg/L.

On day 19, perchlorate was added to the feed media for reactors R2 to R5. Reactor R1 continued with NO_3^- alone, as a control. Figure 5.4 to Figure 5.7 show the perchlorate and nitrate (or oxygen, for R5) results for R2 to R5. The perchlorate removals in all reactors were modest in each reactor after one day of perchlorate addition. The initial removals for R2 and R3, with influents of 100 μg/L and 1000 μg/L, respectively, were 27 μg/L and 36 μg/L, respectively. The initial removal for reactor R4, with an influent of 10,000 μg/L, was around 50 μg/L. Reactor R5, with an influent of 1,000 μg/L, had an initial removal of 45 μg/L.

Over the next few days, reactors R2 and R5 experienced steady improvements in perchlorate removals. In reactor R2, the perchlorate concentration steadily declined and reached non-detect within four days. Reactor R5 also experienced a steady improvement in perchlorate removal until reaching non-detect after approximately 13 days.

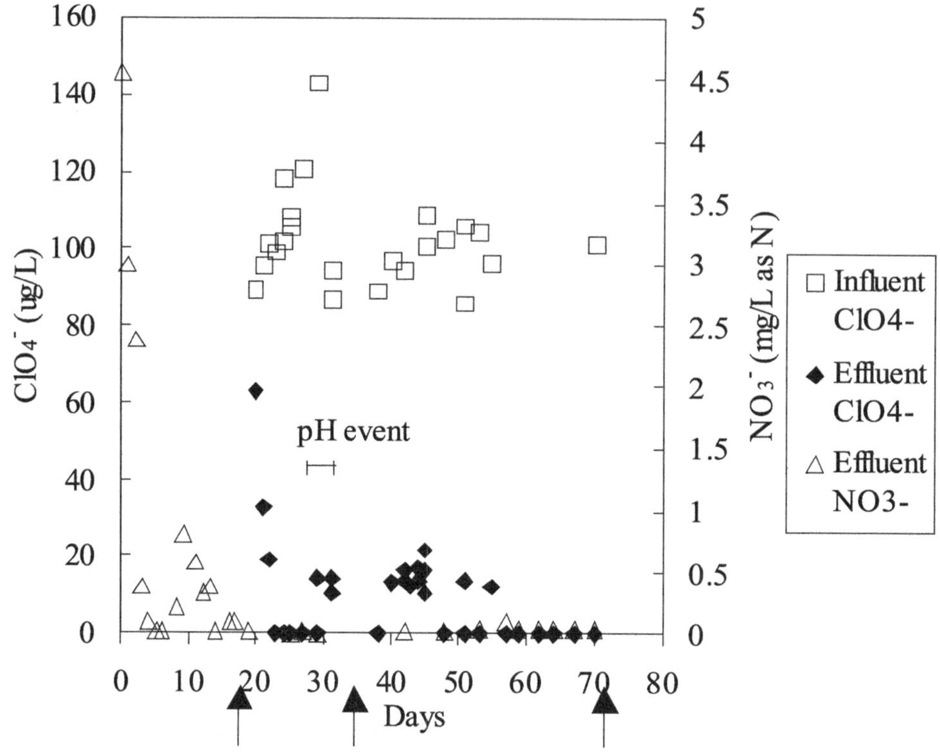

Figure 5.4 Reactor R2 effluent nitrate and influent and effluent perchlorate concentrations (arrows indicate coupon-sampling dates)

On day 28, the amounts of monobasic and dibasic phosphate for the medium for reactors R1 through R4 were accidentally reversed, resulting in an influent pH of 6.2 instead of 7.2 and an effluent pH of 6.8 instead of 7.4. The low pH lasted from day 28 to day 31. While the pH change had little effect on the effluent nitrate concentration, which remained below 15 μg-N/L on these days and thereafter, it appeared to have a significant effect on perchlorate removals. Prior to the pH event, the effluent perchlorate in reactors R3 and R4 was decreasing similarly to that of R5, which was supplied from a separate medium bottle containing 1,000-μg/L perchlorate. However, after the low-pH event, the effluent perchlorate in reactor R4 stopped its decreasing trend, and the effluent perchlorate in R3 began to increase. In reactor R2, the effluent perchlorate appeared to increase to slightly above the detection limit, whereas before the pH event it had been below the detection limit for 8 days.

On day 40, the influent flow rate to reactors R1-R4 was lowered from 1 mL/min to 0.85 mL/min. This caused the effluent perchlorate concentrations to decrease in reactors R2-R4. Table 5.8 shows the influent and effluent perchlorate concentrations and removals at days 36 and 72, when biofilm samples were collected for molecular analyses. The percent of perchlorate removals for reactors R2 and R5 were similar on both sampling dates, and they were substantially higher than for R3 and R4. The percent removals for R3 and R4 were slightly higher on day 71 than on 36.

The relatively poor performance of R3 and R4 after the pH event suggests that the pH change shocked the PCRB, and, although they continued to reduce perchlorate, they were less able to compete with the DNB. These results are parallel to the results of the mechanisms

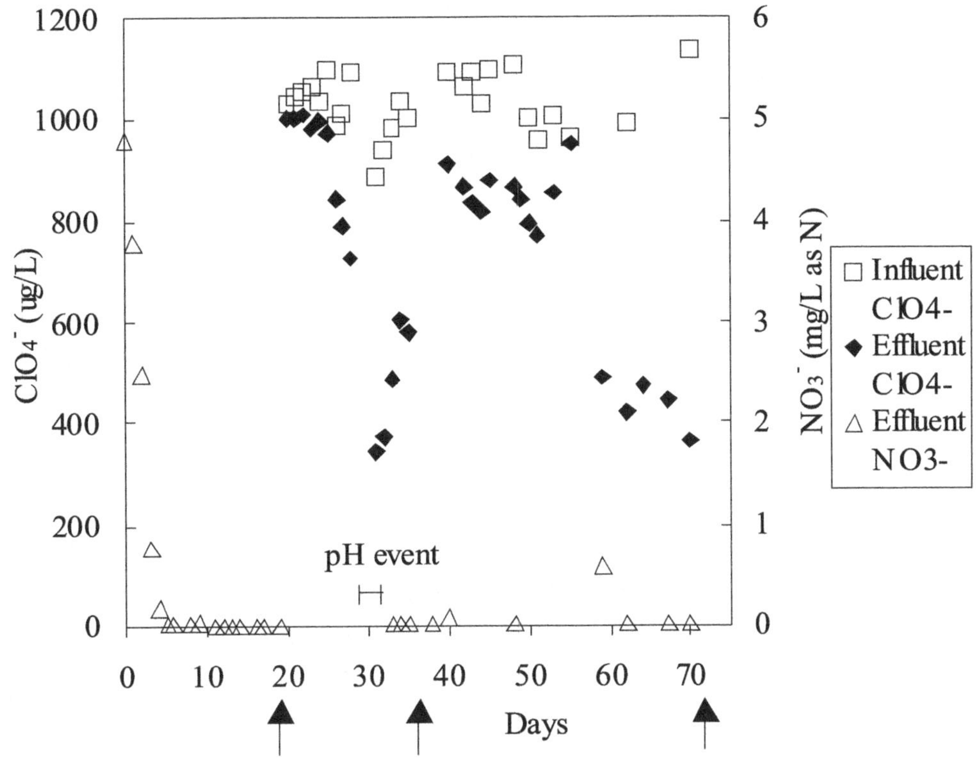

Figure 5.5 Reactor R3 effluent nitrate and influent and effluent perchlorate concentrations (arrows indicate coupon-sampling dates)

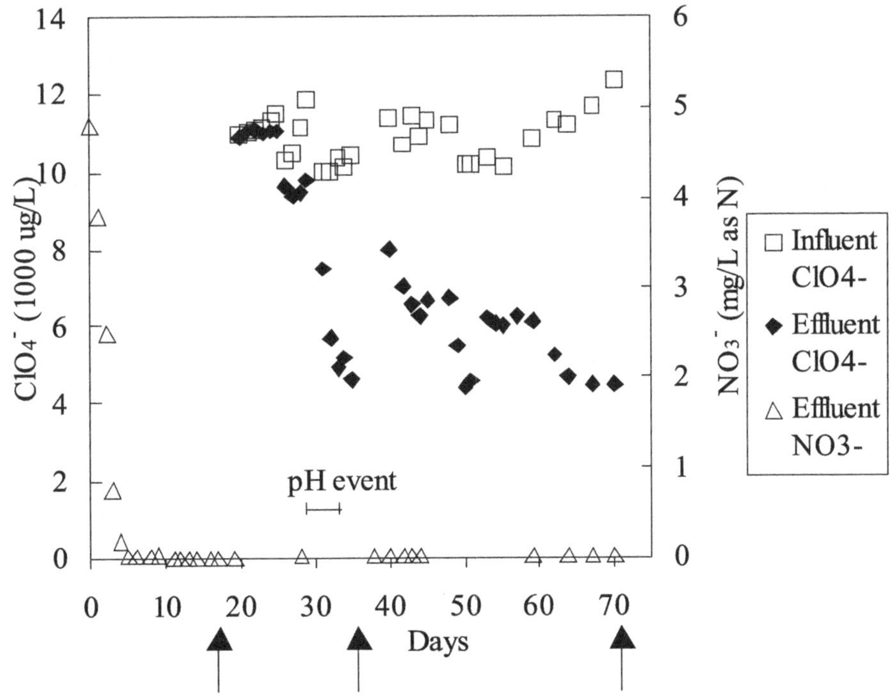

Figure 5.6 Reactor R4 effluent nitrate and influent and effluent perchlorate concentrations (arrows indicate coupon-sampling dates)

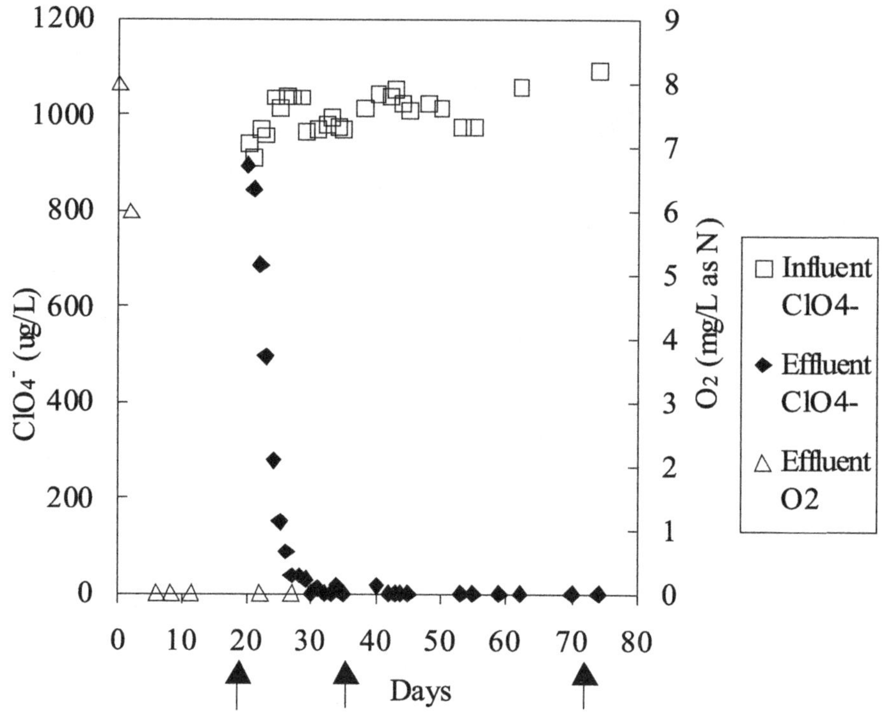

Figure 5.7 Reactor R5 effluent oxygen and influent and effluent perchlorate (arrows indicate coupon-sampling dates)

Table 5.8
Perchlorate concentrations and removals at days 36 and 71

	Day 36				Day 71			
	Influent	Effluent	Removal		Influent	Effluent	Removal	
Reactor	(µg/L)	(µg/L)	(µg/L)	%	(µg/L)	(µg/L)	(µg/L)	%
R2	90	15	75	83	100	<15	>85	>85
R3	1000	580	420	42	1,100	360	740	67
R4	10,400	4,600	5,800	56	11,700	4,500	7,200	62
R5	970	<15	>955	>98	1,100	<15	>1,085	>99

experiments, which showed that perchlorate reduction is more sensitive to pH than nitrate reduction and that unfavorable changes in the medium can cause the reduction or loss of PCRB.

Another important result is that R5, which contained no nitrate in its medium, consistently reduced 1,000-µg/L perchlorate to non-detect. This suggests that, with an appropriate seed, perchlorate reduction can be carried out in waters with no nitrate, as long as they contain dissolved oxygen. This has important practical implications, as some perchlorate-contaminated waters do not have nitrate. It is interesting to note that R5 had much higher removals than R3, even though they had the same influent perchlorate concentration. While it is possible that the pH

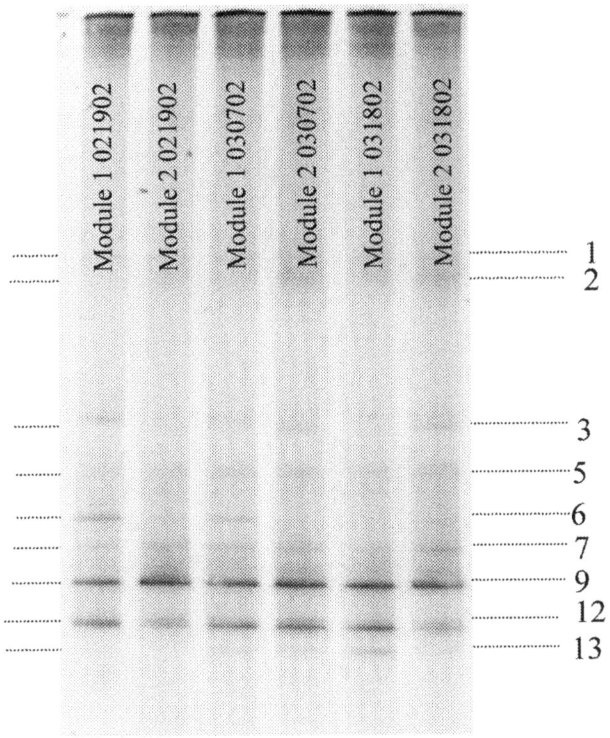

Figure 5.8 DGGE for inoculum biofilm

shock was the sole cause for the poorer performance of R3, it also is possible that nitrate inhibition in R3 decreased the perchlorate-reducing activity of PCRB and thus slowed their enrichment compared to R5.

The perchlorate fluxes on day 71 for reactors R2, R3, R4, and R5 are 14, 125, 1214, and 215 $mgClO_4^-/m^2$-day, respectively. This is based on the data in Table 5.8, a flow rate of 0.85 mL/min for R2-R4 and 1 mL/min for R5, and a fiber surface area of 72.6 cm^2. The nitrate flux for reactors R1 through R4 is 838 mgN/m^2-day.

Molecular Experiments

DGGE

DGGE analyses were performed on the biofilm used to inoculate the reactors and on biofilm collected from reactors R1 through R5 at three times: on day 20, just before perchlorate addition began; on day 36, which was 16 days after perchlorate addition; and on day 71, which was 51 days after perchlorate addition. Figure 5.8 shows DGGE band patterns from the pilot-scale MBfRs from which the inoculum was collected.

The DGGE results for the bench-scale MBfR sampling on day 20 are shown in Figure 5.9. Banding patterns are similar for all five reactors and similar to that of the inoculum biofilm (Figure 5.8). This is expected, as reactors R1-R4 were run with almost the same oxygen and nitrate concentrations from which they were seeded. The two prominant bands in Figure 5.9,

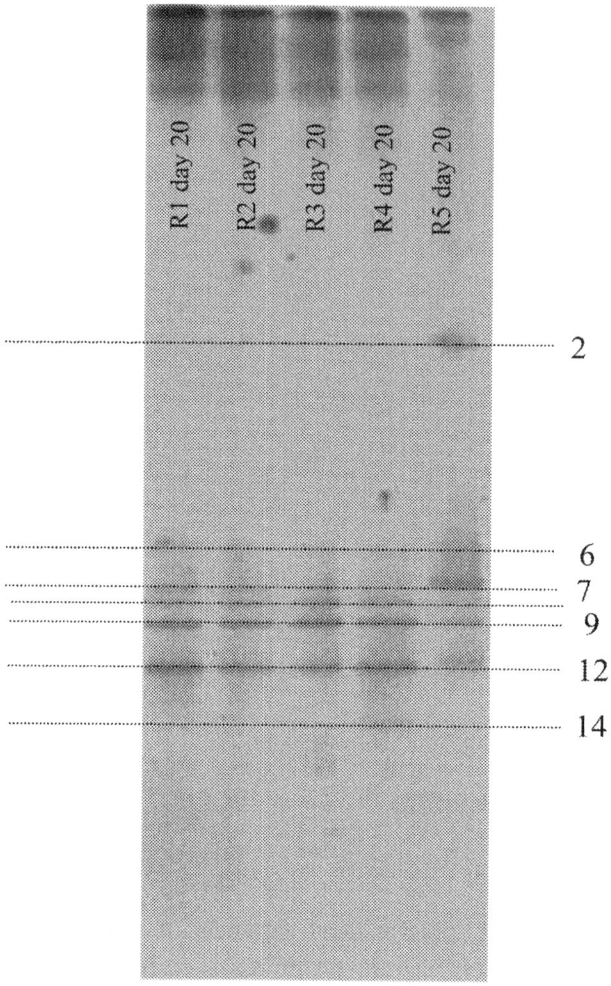

Figure 5.9 DGGE results for sampling on day 20—no perchlorate (all band numbers from all samples are shown in Figure 5.10)

numbers 9 and 12, are in the same position as the main bands in the inoculum (Figure 5.8). Unlike reactors R1-R4, the strongest bands in reactor R5 on day 20 are numbers 2 and 7, although bands 9 and 12 are also clearly present. The general similarity in banding patterns in the denitrifying reactors, R1-R4, and the oxygen reactor, R5, probably occurred because all conditions selected for autotrophic, hydrogen-oxidizing, facultative anaerobes. Additionally, although the reactors were at steady-state removal of nitrate and oxygen at 20 days, the bacteria in the reactor represented the history of the reactor from day 0 to day 20.

Figure 5.10 shows the DGGE results for biofilm collected from reactors R1-R5 for day 36, which is 16 days after perchlorate addition. For reference, the figure also shows selected results from the inoculum biofilm, from reactors R1 and R5 from day 20.

The banding patterns for reactors R1-R5 are slightly different from those of the inoculum and the samples from day 20. On day 36, band 9 has gained more prominence for reactors R1-R4, but band 12 is much less prominent. For R5, band 7 is prominent on day 20, but band 9 also becomes dominant on day 36. By day 36, perchlorate had been supplied to reactors R2-R5 for

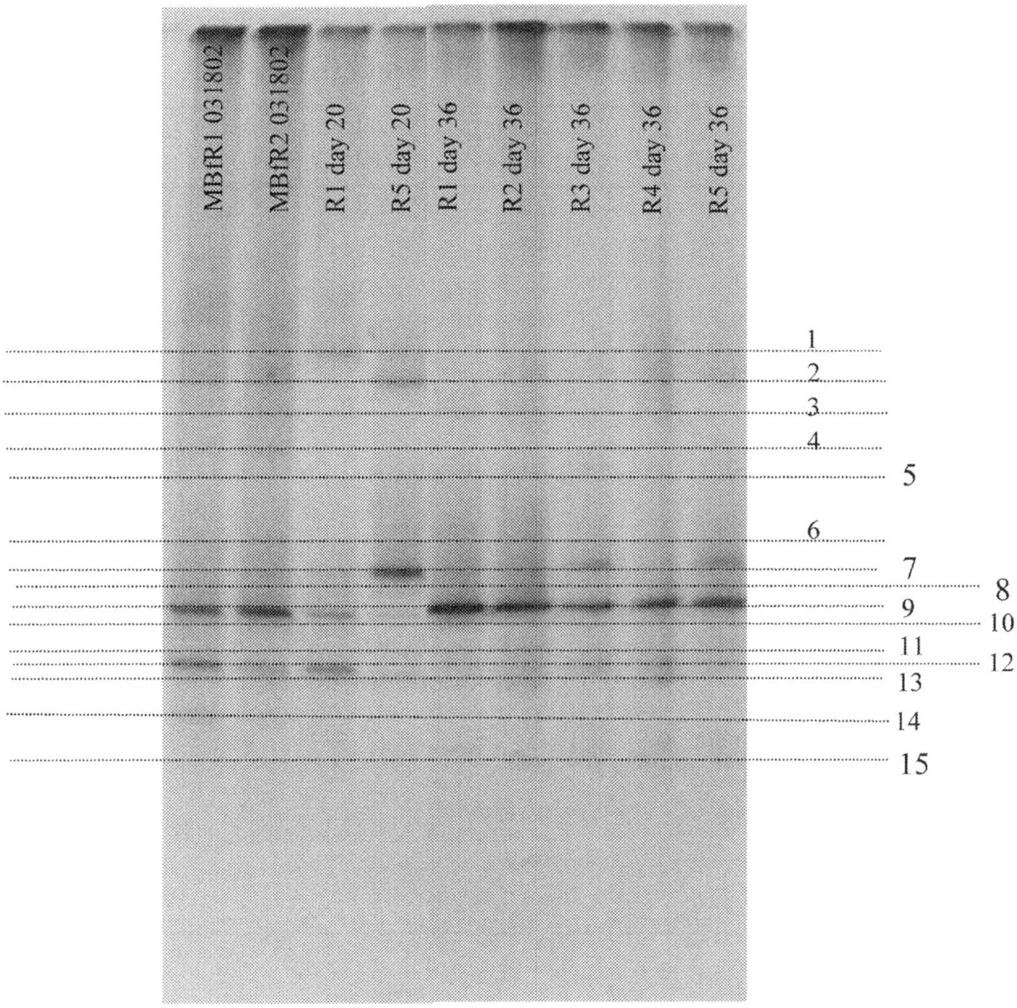

Figure 5.10 DGGE results for day 36, with 16 days of perchlorate addition for reactors R2-R5

16 days. The increasing prominence of band 9 is likely tied to sustained perchlorate input. However, R1 did not receive perchlorate, and it had nearly the same change in band 9 as R2-R5. Therefore, it also is possible that the prominence of band 9 at day 36 was an adaptation to the general conditions in the bench-scale MBfRs.

The bands from the DGGE were extracted and sequenced for day 36. Comparing the sequences to a 16S rDNA library, the percent similarities were tabulated, as shown in Table 5.9. The main bands for the inoculum, numbers 9 and 12, are most similar to *Dechloromonas* JJ (99.6% similarity) and *Rhodocyclus* sp. (95.6% similarity). For day 36, the main bands are 7 and band 9, which are close to *Hydrogenophaga* sp. (96.4%) and, again, *Dechloromonas* JJ.

Figure 5.11 provides the DGGE results for the sampling on day 71, which is after 51 days of continuous perchlorate supply to reactors R2-R5. For reference, the figure also shows selected results for the inoculum samples, reactors R1 and R5 for day 20, and reactors R1, R2, and R5 for day 36.

Table 5.9
DGGE band sequence similarity to known bacteria based on 550 bp partial sequences of 16S rDNA

Band number	Closest match	Percent similarity	Potential metabolism closest match (PCRB or DNB)
1	*Dechloromonas* sp. HZ	92.5	PCRB/DNB (Zhang, Bruns, and Logan, 2002)
2	*Chryseobacterium* sp.	97.1	
3	*Dechloromonas* sp. JJ	95.5	DNB (Coates et al., 2001)
4	*Cytophagales* str.	84.6	
5	Unknown	—	—
6	*Dechloromonas* sp. JJ	96.1	DNB (Coates et al., 2001)
7	*Hydrogenophaga* sp.	96.4	DNB (Aragno and Schlegel, 1992)
8	Unknown	—	—
9	*Dechloromonas* sp. JJ	99.6	DNB (Coates et al., 2001)
10	*Dechloromonas* sp. HZ	87.5	PCRB/DNB (Zhang, Bruns, and Logan, 2002)
11	*Zoogloea*	97.3	
12	*Rhodocyclus*	95.6	
13	*Magnetospirillum* sp.	99.6	
14	*Mycobacterium* sp.	96.7	
15	*Rhodococcus* sp.	99.4	

For day 71, R1 and R2 have prominent bands in the 7, 9, and 12 positions, while reactors R4 and R5 have prominent bands in the 1, 2, 7, and 9 positions. For all five reactors, the most prominent band is in position number 9, which is similar to *Dechloromonas* JJ. The most prominent band for R5 also appears to be band 9. Band 9 has the least intensity for reactor R1, with increasing intensities for bands R2, R3, and R4. The intensity for R5 is intermediate between R1 and R2. Conversely, bands 7 and 12 appear to be strongest for R1 and become weaker for reactors R2 through R5. Bands 1 and 2 are only prominent for reactors R3 and R4.

Although the sequence for band 9 is most similar to *Dechloromonas* JJ, a DNB and not a PCRB, it is important to note that most bacteria in the *Dechloromonas* genus are PCRB, and many are closely related to non-PCRB (Bruce, Achenbach, and Coates, 1999; Coates et al., 1999). For example, based on the 1500 bp 16S rRNA sequence, the PCRB *Dechloromonas* sp. SIUL is 99% similar to *Dechloromonas* JJ, while PCRB *Dechloromonas* sp. PC1 is 98% similar to *Dechloromonas* JJ. Another interesting obsrvation is that the DGGE band for *Dechloromonas* sp. PC1 is in the same position as band 9 (Figure 5.12).

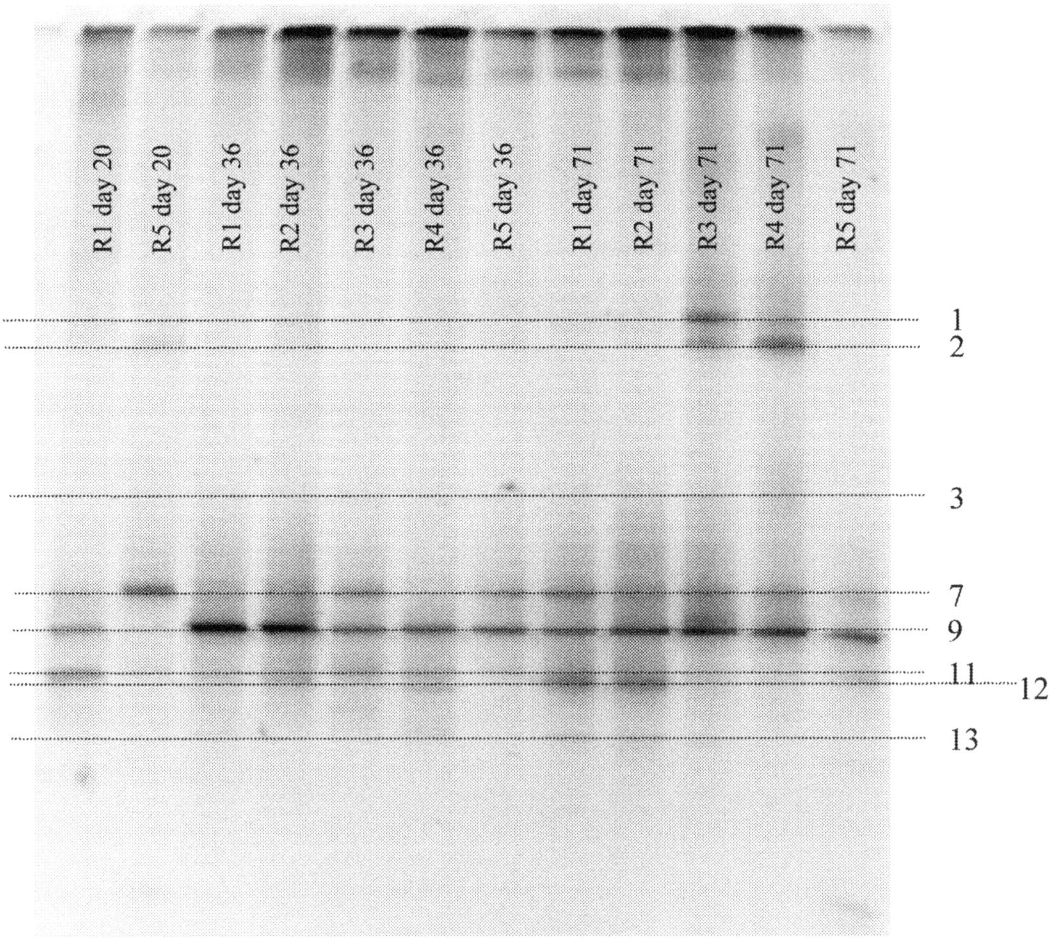

Figure 5.11 DGGE results for day 71 (band numbers refer to position from Figure 5.10)

FISH

The biofilm collected from the sample at day 71 was tested with two FISH probes. One probe, probe PB9, was designed to hybridize with the main band from the DGGE tests (band 9), and the other, probe CH1, was designed to hybridize with strain CH1, a hydrogen-oxidizing, autotrophic bacterium isolated from the biofilm from the pilot-scale reactors. Both probes, together with a general RNA stain, were applied to the reactor biofilm. The results are presented in Figure 5.13.

In reactor R1, which had 5-mg-N/L nitrate and no perchlorate in its medium, 14% of the total bacterial count hybridized with the PB9 probe. In reactor R2, which had 5-mg-N/L nitrate and 100-μg/L perchlorate, the main band bacteria increased to 22%. In reactors R3 and R4, the percentage increased to 31%, and 49%, respectively. In reactor R5, the percentage was only 4%.

Based on the amounts of oxygen, nitrate, and perchlorate consumed and the theoretical yields, Figure 5.14 provides the expected percentages of total biomass on each acceptor for each reactor. Clearly, the actual percentage of PB9 hybridization was much larger than predicted by perchlorate reduction alone when nitrate also was present. Furthermore, PB9 hybridization

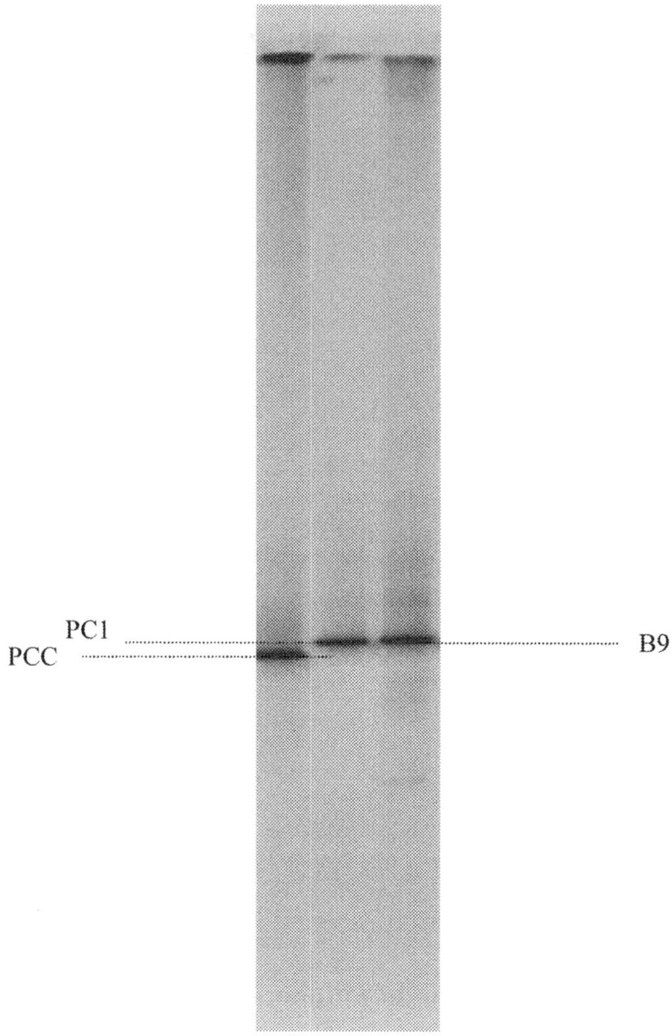

Figure 5.12 DGGE with strains PC1 and PCC and band 9 from day 20

correlated with perchlorate removal when nitrate was present. These two trends suggests that nitrate can select for PCRB, but the reduction of perchlorate provides enrichment beyond what would be expected from the yield on perchlorate.

Perhaps a surprising result is that the abundance of PB9 was only 4% for reactor R5, which is less than for R1, which had no perchlorate. One possible explanation is that band 9 for day 71 was a species with similar, but not identical, sequence to PB9. This is supported by the fact that *Dechloromonas* PC1, a PCRB isolate with a different sequence from PB9, was in the same position as band 9 (Figure 5.12). Thus, a different species of PCRB could have accumulated in R5 in the band 9 position, but with a sequence that did not hybridize with the PB9 probe. Also, the abundance of PB9 in R5 was equal that of CH1, which was also 4%, yet CH1 was not strong enough to produce a detectable DGGE band in Figure 5-10. For day 71, band number 9 for R5 was approximately as strong as for R1 or R2, but these had a PB9 abundance of 14 and 22%, respectively.

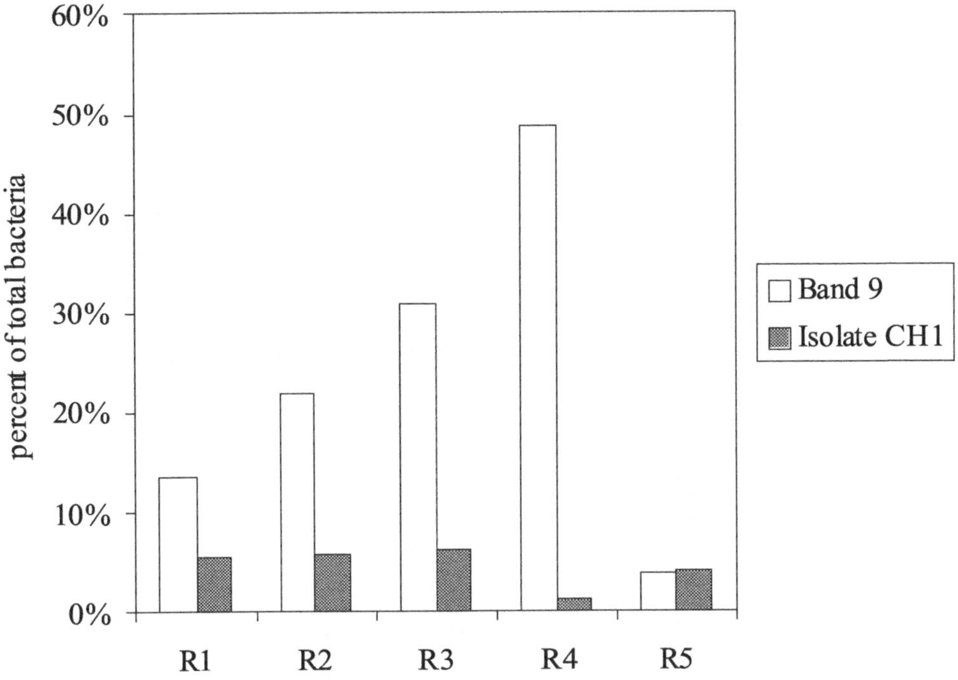

Figure 5.13 FISH results

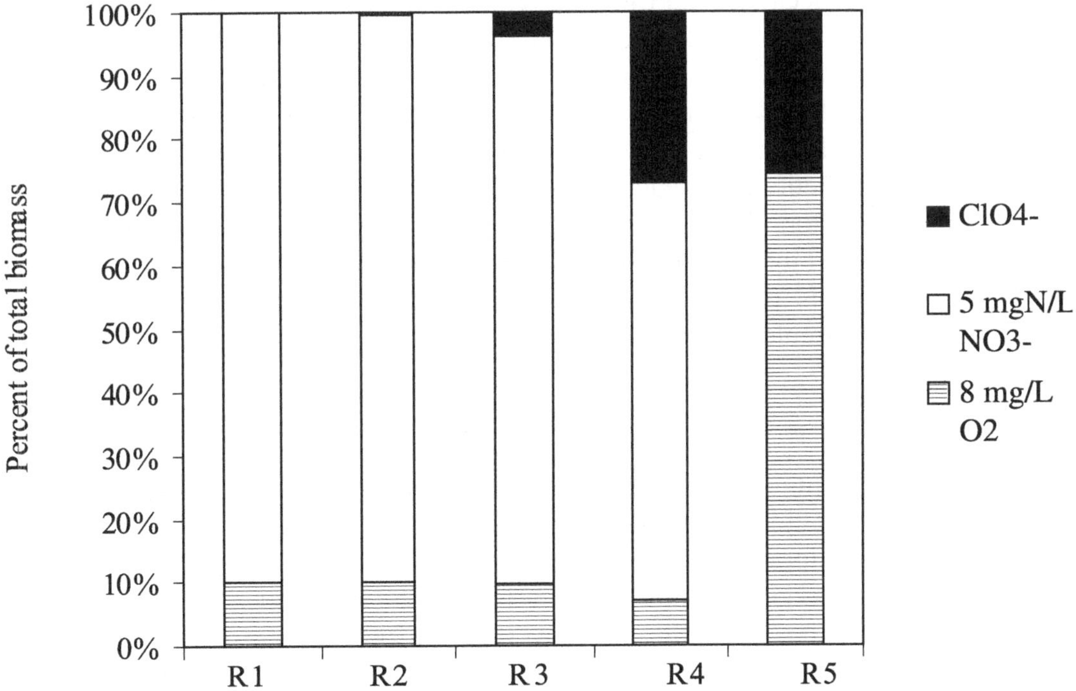

Figure 5.14 Theoretical biomass fractions based on acceptor utilization

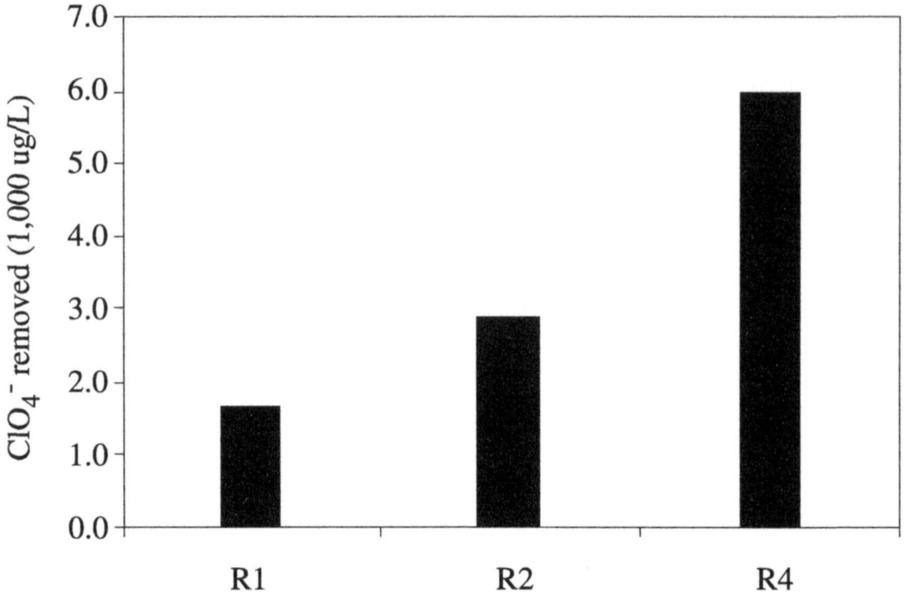

Figure 5.15 Perchlorate removals with a 10,000 µg/L influent in the activity tests

Activity Test

An activity test was carried out to further explore the ecology of the MBfRs. Reactors 1, 2, and 4 were challenged with a medium containing 10,000-µg/L perchlorate and 5-mg/L nitrate. The test was carried out around 20 days after the last coupon sampling. The average removals at the end of the tests are plotted in Figure 5.15. R3 and R5 were not operating at the time. These results agree with the FISH results, which suggest that a key PCRB, PB9, was enriched by sustained perchlorate reduction. For reactor R1, the abundance of PB9 by FISH hybridization was 14 percent, and the perchlorate activity was 1,700 µg/L; for R2, the abundance was 22 percent, and the activity was 2,900 µg/L; and in R4 the abundance was 49 percent, and the activity was 5,900 µg/L. Thus, PB9 is present in the reactor without any perchlorate, but increases in abundance with higher perchlorate reduction.

Conclusions

The experiments highlight several important findings about the microbial ecology of MBfRs that can reduce perchlorate. First, oxygen can serve as a primary substrate that selects for PCRB. In fact, the perchlorate removal in the oxygen reactor was greater than in the nitrate reactor, suggesting that oxygen may be a preferred primary acceptor. The advantage of oxygen may be explained by that fact that nitrate inhibits perchlorate reduction, even at low concentrations, as shown in the mechanisms experiments presented earlier in this report. Being able to use oxygen as a primary acceptor has important practical advantages, especially for treating perchlorate-contaminated groundwaters without nitrate.

A second finding is that, although PCRB were present in the reactor without perchlorate, a PCRB, strain PB9 was enriched by the presence of perchlorate. These results also confirm the results presented earlier in this report, which suggest that the presence of perchlorate enriched for

PCRB. Even a small influent concentration of perchlorate (100 μg/L) provided a significant enrichment for PB9 and increased the reactors perchlorate-removal capacity. The increase in PB9 abundance was much greater than expected from the yield on perchlorate alone when nitrate was present, suggesting that PB9 gains an advantage in competing for nitrate when perchlorate is present. Although PB9 appeared to be the dominant PCRB in the reactors receiving nitrate, CH1 also was present, and other PCRB species could have been present as well. In particular, PC1 or another closely related PCRB could have been important in R5, which received no nitrate.

The third finding is that a number of autotrophic, denitrifying PCRB, including CH1, PC2, and CA1, were isolated from the inoculum. This suggests that autotrophic perchlorate reduction with hydrogen under denitrifying conditions is not restricted to one or a small number of bacteria. It is likely that such bacteria are diverse and common in the environment and that an MBfR should develop an appropriate perchlorate-reducing community without any special inoculation.

CHAPTER 6
MEMBRANE BIOFILM REACTOR DEMONSTRATION

Based on the knowledge developed from the initial feasibility investigation, a larger scale MBfR system was designed and constructed with the intent to:

- Determine system operational and design parameters that affect perchlorate biodegradation
- Identify the long-term operational and maintenance issues of a full-scale MBfR for perchlorate removal
- Evaluate reactor designs and the impact of hydraulics, cleaning, and maintenance
- Characterize the impacts of the unit operations downstream of the MBfR that would be required for the production of potable water

To accomplish the objectives of this research, suites of analytical methods, experimental systems, and procedures were employed. The analytical methods were used to assess the water quality and quantify the performance of the experimental systems. Experimental procedures were used to operationally characterize the microbial ecology of mixed cultures. During the course of the research, the methods and procedures were subjected to stringent quality control to identify variability or error within the analytical and experimental results.

MATERIALS AND METHODS

Pilot Plant

Designed, constructed, and installed in MWH's Mobile Water Treatment Pilot Trailer during the initial periods of this project, the MBfR pilot plant was located in La Puente in southern California. This site, owned and operated by the La Puente Valley County Water District, has an active groundwater well and employs a full-scale 2,500 gpm Calgon Carbon Corporation ISEP ion-exchange system for the removal of perchlorate. A photo of the exterior of the trailer and system feed tank is shown in Figure 6.1.

Process Description

Figure 6.2 shows a picture of the pilot plant. The pilot plant included various novel hydrogen-fed hollow fiber membrane biofilm reactors (MBfRs), followed by an aeration basin and a granular media filter. The membrane biofilm reactors (Mitsubishi, Japan; Membrana, Charlotte, NC) contained fibers potted either at one or both ends of a cylindrical reactor. The water passing through each individual reactor was recirculated to control the linear velocity through the modules. For each system, an air scour was also periodically applied to the MBfRs to help keep the fibers from sticking together and reduce the clumping of biomass on the membrane surface.

Hydrogen was fed to one end of the reactor, filled the inside of the fibers and diffused through the membranes to serve as an electron donor for the biofilm. The hydrogen pressure was maintained well below the bubble-point of the membrane, eliminating the formation of a hydrogen atmosphere within the bioreactors. The perchlorate-contaminated water was then treated as it passed along the biofilm on the outside of the fibers.

Figure 6.1 Pilot plant trailer and influent reservoir

Figure 6.2 Biological perchlorate reduction pilot plant

Following the MBfRs, an aeration process was used to achieve two primary goals: first, it oxygenated the water in preparation for its introduction into a distribution system as a drinking water source; and second, it provided sufficient oxygen for operating the downstream filter in an aerobic biodegradation mode to achieve complete removal of any residual dissolved hydrogen in the water. A process schematic of this system is shown in Figure 6.3 to help illustrate some of these details.

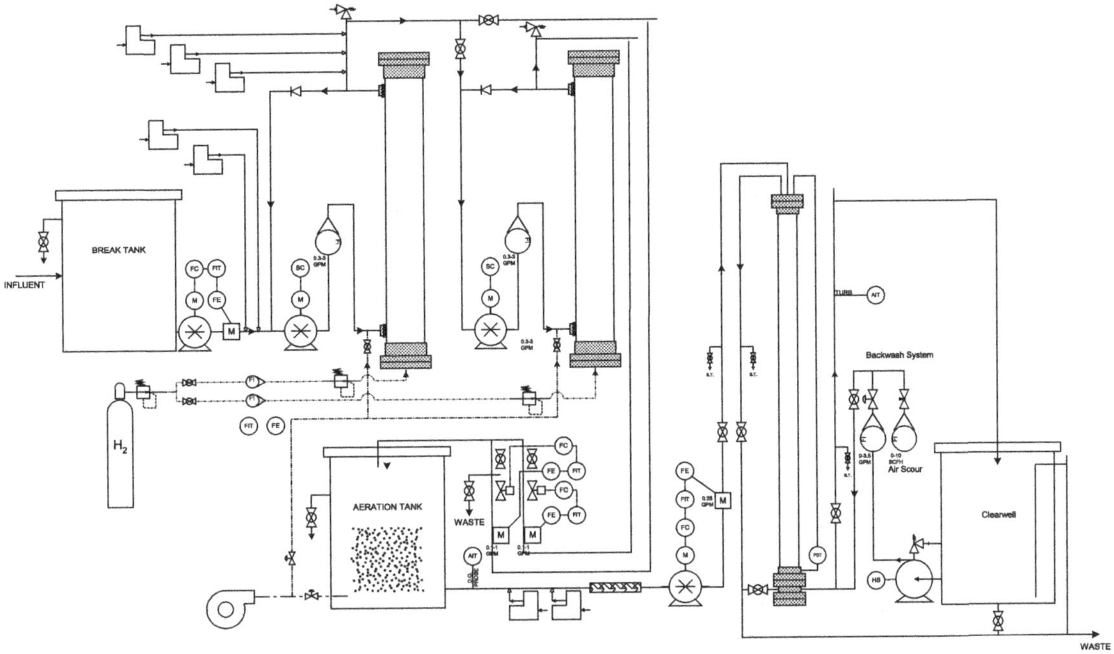

Figure 6.3 Biological perchlorate reduction pilot plant process schematic

Operational and Design Parameters

The MBfRs designed or selected for use in the pilot-plant were full-scale hollow-fiber membrane modules. The operational and design parameters of these reactors, as well as the rest of the system, are summarized in Table 6.1 and Table 6.2.

Process Sampling

The pilot was designed so that samples could be collected from the break tank (influent water), influent and effluent of each bioreactor, aeration tank effluent, and media filter effluent. Sampling sites were selected to provide a complete analysis of a variety of processes used in this study. The frequency of sample collection was based on operational conditions and historical performance.

Influent Water Quality

Selected parameters were monitored in the influent water quality. The influent groundwater was stable for most of the parameters monitored, as summarized in Table 6.3. The alkalinity, temperature, and pH of the water during pilot testing are shown in Figure 6.4. The concentrations of other selected water quality parameters (perchlorate, sulfate, and nitrate) are displayed in Figure 6.5. Apparent from this figure is a decreasing trend in the perchlorate concentration from the beginning of the study. While the well from which the water was drawn from did not operate on a continuous basis, it did not appear to be affecting the water quality.

Table 6.1
Pilot reactor designs and operational parameters

Parameter	Mitsubishi (M3)	Membrana	Mitsubishi (M20)	Mitsubishi (M6)
Process				
Membrane surface area	14.6 m^2	45 m^2	48.4 m^2	14.6 m^2
Bioreactor module				
Number of modules	2	1	1	1
Length	110 cm	50 cm	110 cm	110 cm
Diameter	14 cm	15.2 cm	14 cm	14 cm
Volume	16.4 L	6.9 L	13.5 L	15.9 L
Membrane fiber				
Outside diameter	280 μm	300 μm	280 μm	280 μm
Active length	110 cm	50 cm	110 cm	110 cm
Cross sectional area	0.00062 cm^2	0.00071 cm^2	0.00062 cm^2	0.00062 cm^2
Number/module	8000	62691	50000	15000
Packing density	3 percent	27.3 percent	20 percent	6 percent

Table 6.2
Downstream unit operations design and operational parameters

Parameter	Value
Aeration tank	
Volume	5 gal
Filter column	
Diameter	3 in
Filter loading rate	5 gpm/sq-ft
Media depth	24 in GAC; 12 in sand

Table 6.3
Influent water quality summary

Parameter	Unit	Median	Average	Standard Deviation	Range
Temperature	°C	20.6	21.7	2.86	15 - 26.7
Turbidity	NTU	0.4	0.4	0.08	0.31 - 0.43
pH		8.1	8.1	0.15	7.68 - 8.97
Total alkalinity	mg $CaCO_3$/L	182	181	11.5	120 - 208
Conductivity	µS	491	491.0	2.8	489 - 493
Perchlorate	µg/L	55	54	6.2	40 - 71
Nitrate as N	mg/L	5.9	5.7	0.42	4.3 - 6.9
Sulfate	mg/L	30.6	31	3.7	24.8 - 39.7

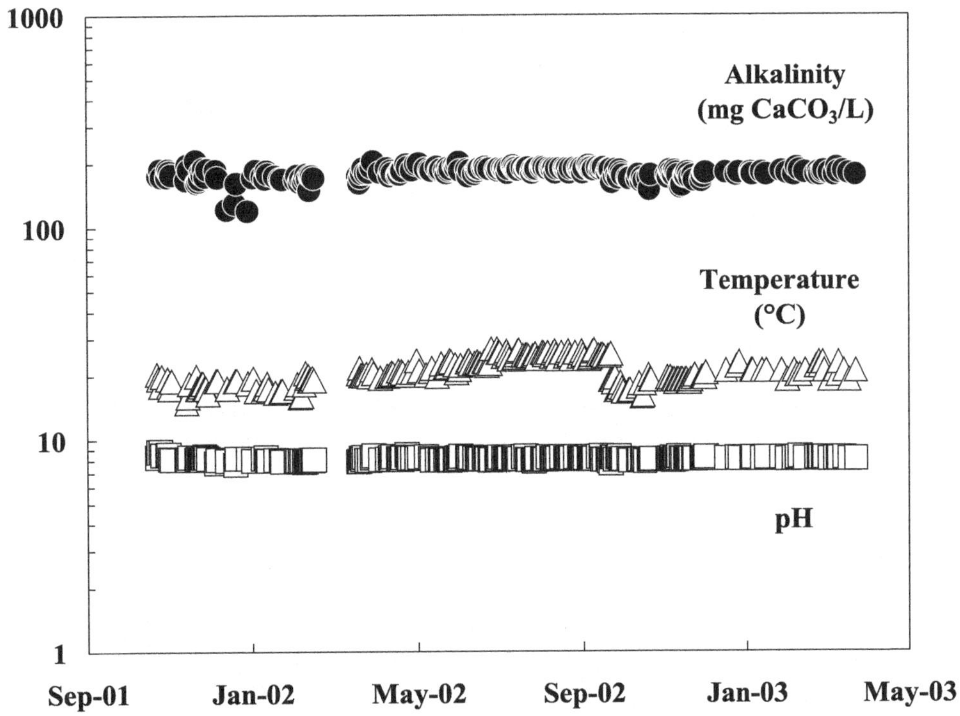

Figure 6.4 General influent water quality (alkalinity, temperature, pH)

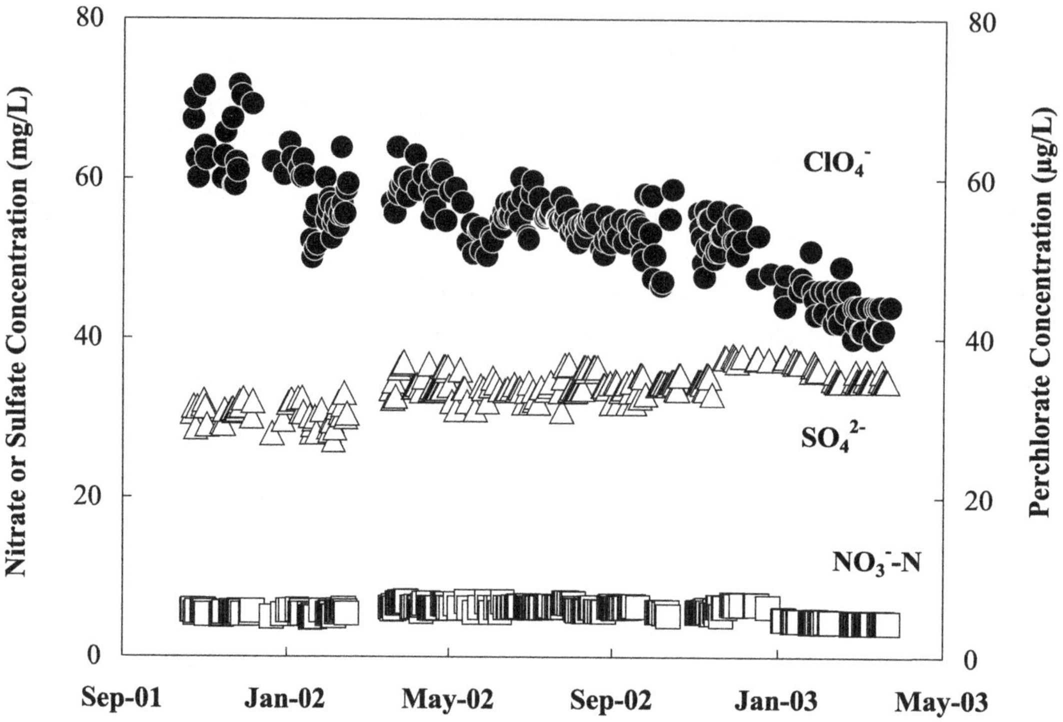

Figure 6.5 Selected influent water quality parameters

Analytical Methods

Most of the water quality parameters were measured following *Standard Methods* (APHA, AWWA, and WEF, 1998) or USEPA methods. A summary of analytical procedures used, both approved and other validated standard methods, is provided in Table 6.4. A discussion of modified and non-standard methods is also included.

Ion Chromatography

Utilizing the latest technology at pilot-scale, perchlorate was analyzed on-site by ion chromatography (IC) using a Dionex DX-320 with conductivity detection using an AS-16 column, a 1000-∝L loop, EG-40 eluent generator, and an autosampler as shown in Figure 6.6. EPA Method 314.0 was followed as the analytical protocol. Based on seven injections of a 2-μg/L standard over the period of one week, the MRL was determined to be 0.7 μg/L. The lowest standard used during calibration was 2 μg/L.

All anions other than perchlorate (chloride, chlorate, chlorite, nitrate, nitrite, and sulfate) were analyzed on the same system using an AS-17 column, a 10-μg/L loop, and an EG-40 eluent generator configured for gradient analysis. Complete resolution of all of these anions of interest was evaluated in a single run using the EG-40 to produce a hydroxide eluent gradient based on EPA Method 300.1 modified for use with a hydroxide-selective column (Jackson et al., 2000).

Table 6.4
Summary of approved or standard analytical procedures

Parameter	Method number	Method title	Reference
ClO_4^-	EPA Method 314.0*	Determination of Low Concentrations of Perchlorate in Drinking Water Using Ion Chromatography	USEPA (1999)
	Modified EPA Method 300.1†	Analysis of Low Concentrations of Perchlorate in Drinking Water and Groundwater by Ion Chromatography	Dionex
ClO_3^- , ClO_2^-	Modified EPA Method 300.1*	Determination of Inorganic Anions in Environmental Waters with a Hydroxide-selective column	Jackson et al. (2000)
	EPA Method 300.1†	Determination of Inorganic Anions in Drinking Water by Ion Chromatography	USEPA (1993)
Cl^-	Modified EPA Method 300.1*	Determination of Inorganic Anions in Environmental Waters with a Hydroxide-selective column	Jackson et al. (2000)
	4500-Cl^- F†	Chloride-Ion Chromatography Method	*Standard Methods* (APHA, AWWA, and WEF, 1998)
NO_3^-	Modified EPA Method 300.1*	Determination of Inorganic Anions in Environmental Waters with a Hydroxide-selective column	Jackson et al. (2000)
	4500-NO_3-C†	Ion Chromatographic Method	*Standard Methods* (APHA, AWWA, and WEF, 1998)
NO_2^-	Modified EPA Method 300.1*	Determination of Inorganic Anions in Environmental Waters with a Hydroxide-selective column	Jackson et al. (2000)
	4500-NO_2-C†	Ion Chromatographic Method	*Standard Methods* (APHA, AWWA, and WEF, 1998)
SO_4^{2-}	Modified EPA Method 300.1†	Determination of Inorganic Anions in Environmental Waters with a Hydroxide-selective column	Jackson et al. (2000)
H_2 (g)	Non-standard method	Orbisphere* Trace Analytical†	
Acetate	4110	Determination of Anions by Ion Chromatography	*Standard Methods* (APHA, AWWA, and WEF, 1998)
Methanol	Non-standard method		
TOC/DOC	5310C	Persulfate Ultraviolet Method	*Standard Methods* (APHA, AWWA, and WEF, 1998)
DO	4500-O G.	Membrane Electrode Method	*Standard Methods* (APHA, AWWA, and WEF, 1998)
Ca^{2+}	Method 200.7	ICP	Hauman and Munch (1997)
Mg^{2+}	Method 200.6	ICP	Hauman and Munch (1997)
Total Hardness	2340C	EDTA Titrimetric Method	*Standard Methods* (APHA, AWWA, and WEF, 1998)
Bicarbonate	4500-CO_2 D.	Calculation	*Standard Methods* (APHA, AWWA, and WEF, 1998)
NH_3	4500-NH_3 F.	Phenate Method	*Standard Methods* (APHA, AWWA, and WEF, 1998)
Total Alkalinity	2320B	Titration Method	*Standard Methods* (APHA, AWWA, and WEF, 1998)

(continued)

Table 6.4 (Continued)

Parameter	Method number	Method title	Reference
Turbidity	2130B	Nephelometric Method	*Standard Methods* (APHA, AWWA, and WEF, 1998)
Place counts	9215B	Pour Plate Method	*Standard Methods* (APHA, AWWA, and WEF, 1998)
Temperature	2550B	Laboratory and Field Methods	*Standard Methods* (APHA, AWWA, and WEF, 1998)
pH	$4500H^+$	Electrometric Method	*Standard Methods* (APHA, AWWA, and WEF, 1998)
Specific Conductivity	2510B	Laboratory Method	*Standard Methods* (APHA, AWWA, and WEF, 1998)

* Pilot-scale
† Bench-scale

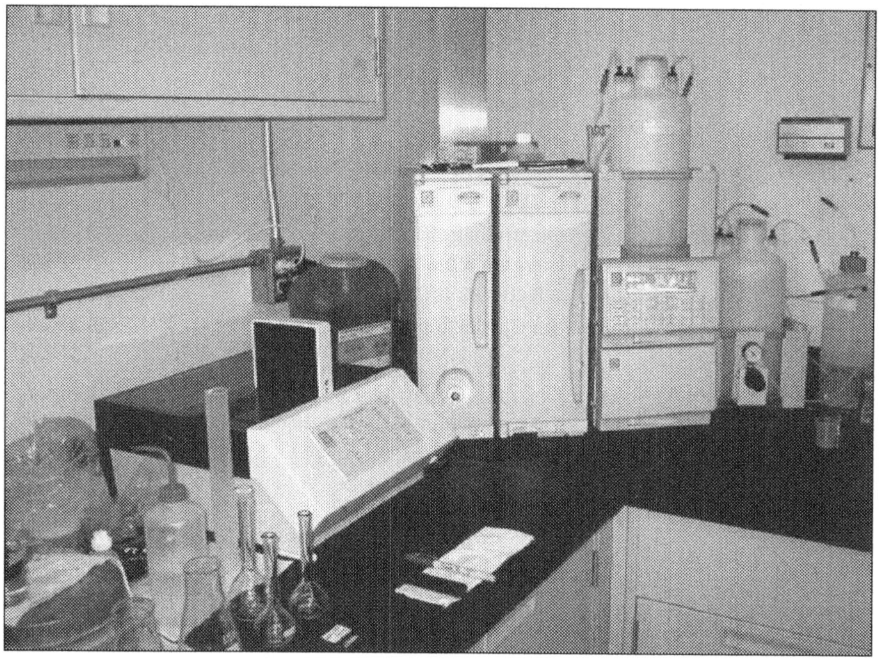

Figure 6.6 On-site ion chromatograph

Hydrogen Analysis

There are no standard methods for analyzing hydrogen. Consequently, dissolved H_2 at the pilot-scale system was analyzed directly using an Orbisphere Model 3654 Portable Micro Logger configured for dissolved hydrogen, as shown in Figure 6.7. Direct measurement of dissolved hydrogen concentrations in the bulk liquid were made utilizing membrane-covered dynamic thermal conductivity (MDTC) sensor technology. The MDTC worked by allowing hydrogen gas to permeate through a hydrogen selective membrane and was measured by the increase in thermal conductivity against a reference gas, nitrogen.

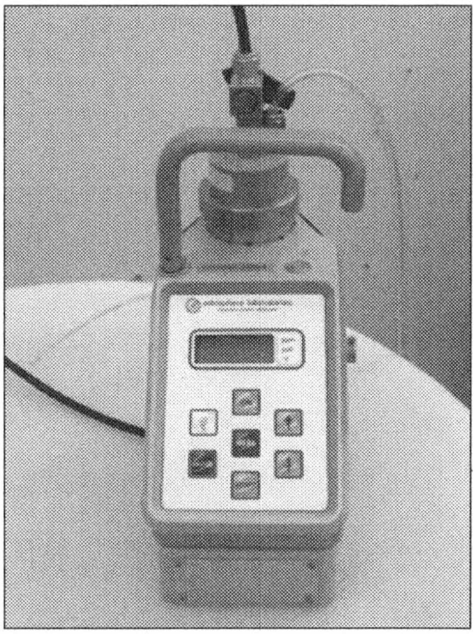

Figure 6.7 On-site dissolved hydrogen analyzer

Quality Assurance/Quality Control

Quality assurance (QA) was maintained through the continued use of the approved Quality Assurance Project Plan (QAPP). The objective of this plan was to ensure that the results and conclusions derived from the project are based on valid data of known and documented quality. The QAPP consisted of the following major elements: 1) quality control (QC), which includes sampling procedures, sample chain of custody, measurement of data, calibration procedures and frequency, and data reduction, validation, and reporting; 2) quality assessment, which includes performance and system audits, and correction action; and 3) reporting.

Project QC was maintained to control analytical measurement errors and included the following elements: sampling procedures, sample chain of custody, development and strict adherence to principles of good laboratory practice (using qualified staff and reliable, well-maintained equipment), verified and specified analytical methods for measurements (including sensitivity, selectivity, precision, and accuracy), and consistent use of standard instrument operation procedures (including calibration and standardization procedures, minimum reporting limits (MRLs) and/or method detection limits (MDLs), and frequency). Project QC resulted in a measurement system operating in a state of statistical control whereby errors are reduced to acceptable levels.

The data management system involved the use of computer spreadsheets as well as a protocol for the entry, verification, and presentation of the data in appropriate tabular and graphical forms. Field data (lab evaluations and field sampling) was recorded by hand on specially prepared data sheets for all operational and water quality measurements and/or observations. These data were entered into the appropriate spreadsheet. Following entry, the spreadsheet was printed out and the hard copy checked against the handwritten datasheet. Any corrections were noted on the hard copy; corrected on the screen and saved followed by a print out of the updated spreadsheet. Data was exported directly from the spreadsheets for use in graphics, statistics, and water quality models.

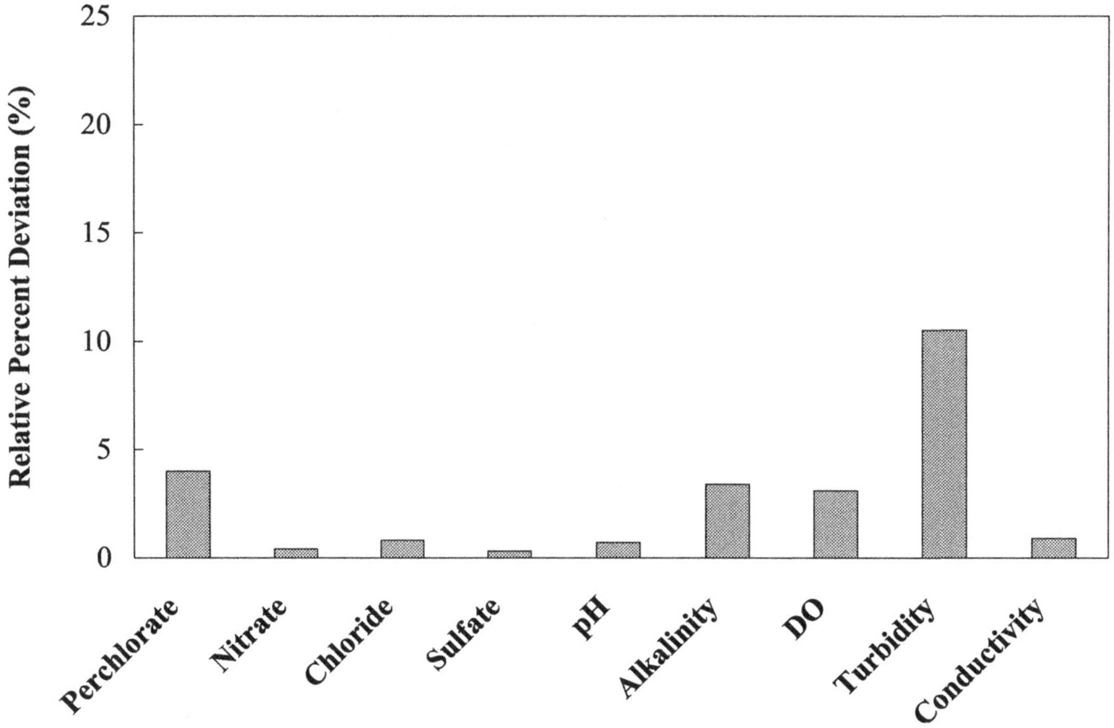

Figure 6.8 Summary of pilot testing QA/QC

As specified in the QAPP, the overall precision of water quality analysis performed by the project team was verified by replicate analysis and calculation of the Relative Percent Deviation (RPD); the goal was defined to be less than or equal to 25 percent. As seen in Figure 6.8, the QA goal was consistently met for all the analyses with the majority of analysis having an RPD of less than 5 percent. Although still acceptable compared to the QA goal, turbidity analysis had a RPD of 10 percent due to the extremely low levels of turbidity observed.

PILOT-SCALE TWO REACTORS IN SERIES (M3)

Start-up

The goal of the start-up period was to operate the pilot system to encourage the initial development of a perchlorate-reducing biofilm. The system was started at a flow rate of 0.10 gpm. Over the course of a four-week period, the system flowrate was increased from 0.10 gpm to the design flowrate of 0.30 gpm. The low flowrate of 0.10 gpm was initially used to allow sufficient residence time in the bioreactors to promote the attachment of indigenous groundwater bacteria to the fibers. Additionally, no recycle was used during these first four weeks to minimize the linear velocity along the fibers and reduce the potential shear of attached organisms. It is important to note that other than the indigenous organisms already present in the groundwater, the MBfRs were not seeded with organisms.

Figure 6.9 reveals that, by evaluating the performance of the system using a normalized substrate flux based on the membrane surface area of the entire MBfR system, the total mass of

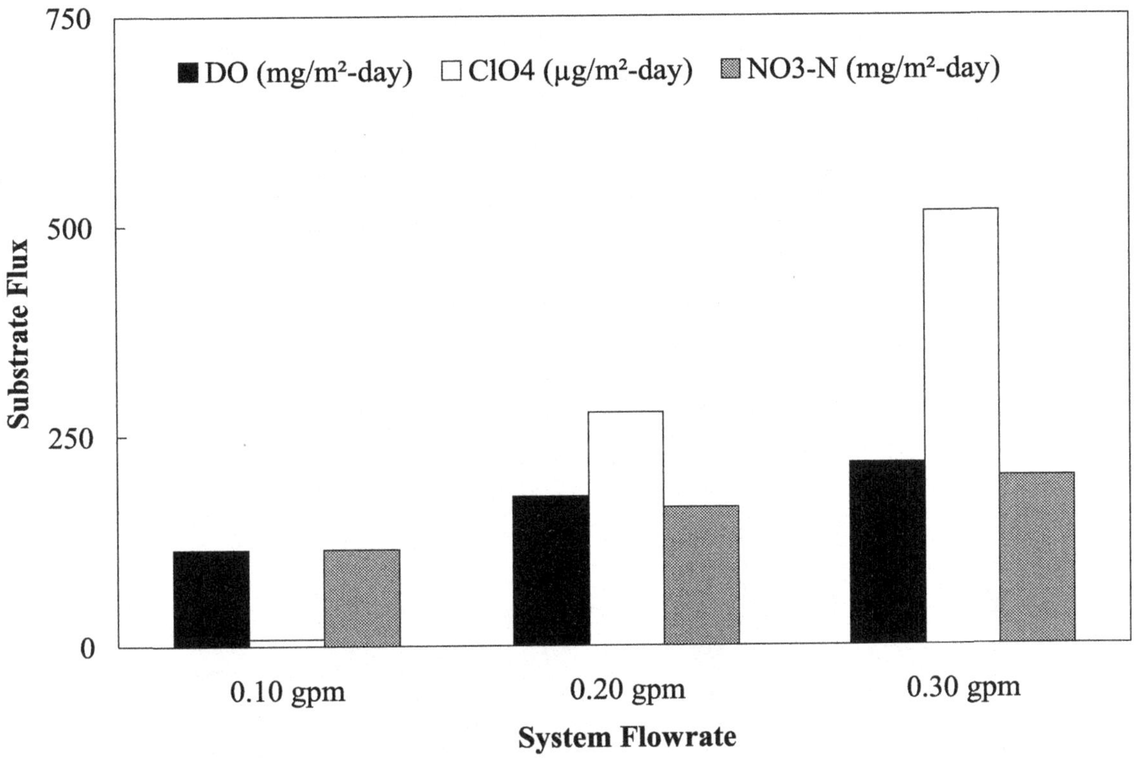

Figure 6.9 Initial MBfR performance (substrate flux)

nitrate and DO reduced nearly doubled during this first four-week period. The increased substrate fluxes at each flowrate exhibited the successful growth and development of a biofilm capable of reducing DO, nitrate, and perchlorate.

Biofilm Development

After the biofilm development had been initiated during the start-up period, recycle flow within each reactor began. The recycle flow was used to foster the development of a denser, more robust biofilm, and to encourage the even distribution of the nutrients amongst the fibers in each reactor. At the system flowrate of 0.30 gpm, each reactor was operated with a recycle flow rate of 1.8 gpm (R=6) for approximately three weeks. For the following two months, the system operated with a recycle ratio equal to 15. Further increases of the recycle ratio to 25 and 50 followed this two-month period. Each increase in the recycle flowrate was a response to an observed plateau in perchlorate removal.

As the pilot testing continued, the biofilm development was visibly noticeable. Figure 6.10 shows the development of this growth. The membrane fibers of the new reactor started out having a brilliant white color and being free of any biofilm. Prior to starting the recycle flows in each reactor, discrete patches of fluffy biomass were attached to some of the fibers. During the biofilm development stage, the biofilm grew to cover most of the fibers and developed a beige color until a dense biofilm completely and uniformly covered the membrane fibers.

A hydraulic test was performed near the end of the biofilm development stage to evaluate the development of the biofilm. For this test, the recycle in each reactor was temporarily shut off

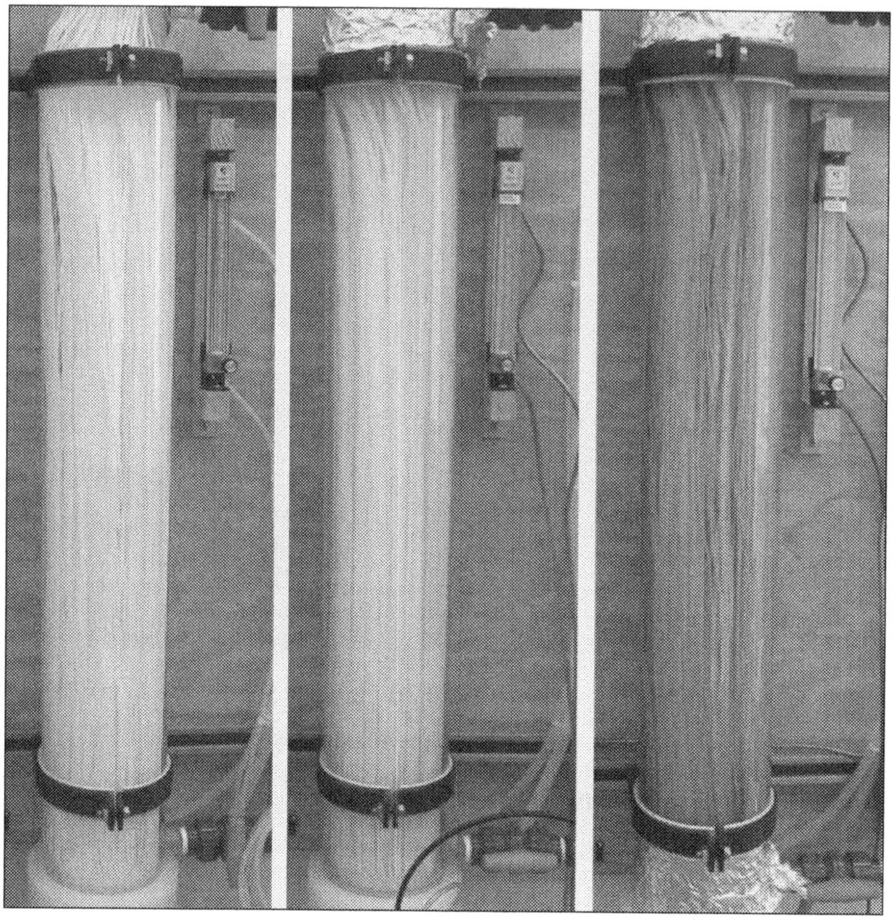

Figure 6.10 Visual development of biofilm growth

for four theoretical hydraulic residence times. Comparing the test results to some of the data collected during the start-up period, Figure 6.11 shows a noticeable improvement. Perchlorate reduction more than tripled compared to the performance observed during the start-up period. This improvement demonstrated continued biofilm growth and confirmed that the improved performance was not solely attributable to the change in hydraulics as a result of recycling water within each reactor.

Steady-State Operation (2 reactors in series)

At steady-state, the MBfR system operated at a flowrate of 0.15 gpm, corresponding to a theoretical hydraulic residence time of 56 minutes. The water within each of the two reactors in series was separately recycled at 4.0 gpm, resulting in a recycle ratio (R) of 27 and an average linear velocity of 100 cm/min (1.7 cm/sec) through each reactor. The hydraulic pressure in the two reactors was 35 psig and 25 psig, respectively. Hydrogen was delivered to the biofilm through the lumens of the membrane fibers at a rate 3.9 mg/min.

Under the operating conditions identified above, the system performance was measured at 96, >97, and >99 percent removal of perchlorate, nitrate, and oxygen, respectively (see Table 6.5).

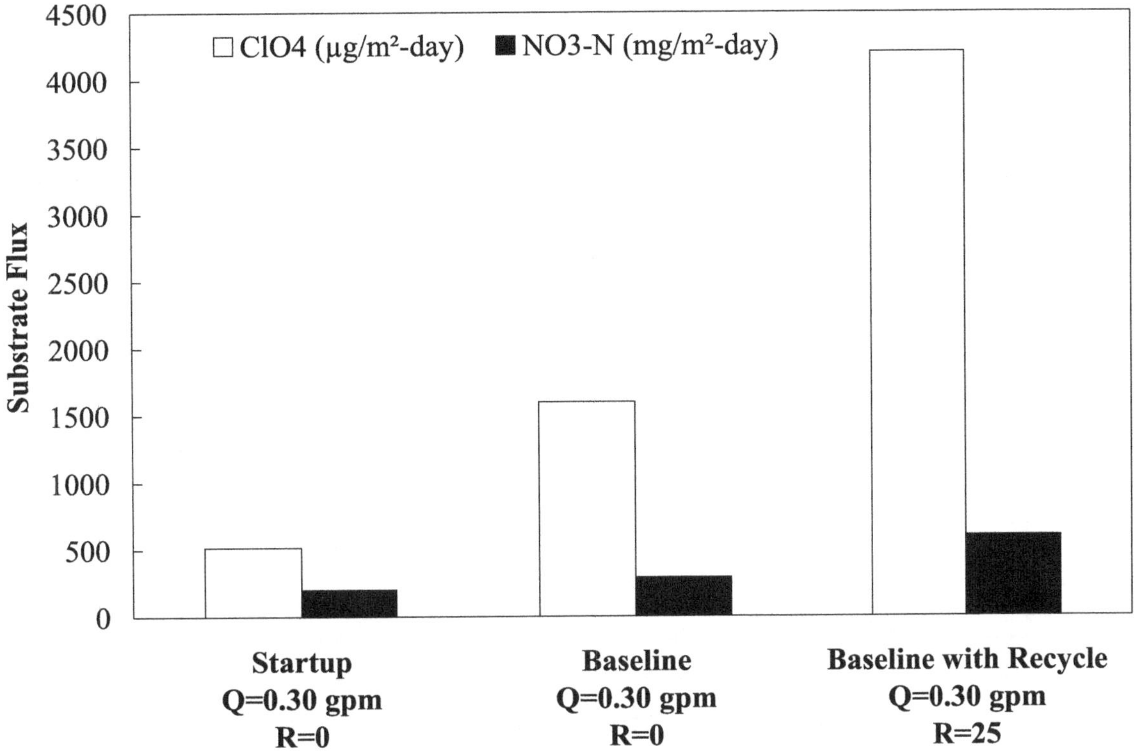

Figure 6.11 Influence of biofilm growth on substrate flux

Table 6.5
Reduction of oxygen, nitrate, and perchlorate

Parameter	Effluent concentration	Removal efficiency	Substrate flux
Perchlorate	2 µg/L	96%	3250 µg/m^2-day
Nitrate	<0.02 mg-N/L	>97%	671 mg-N/m^2-day*
DO	< 0.10 mg/L	>99%	860 mg/m^2-day*

* Substrate flux calculated based on complete reduction of nitrate and oxygen in the first reactor only (membrane surface area per reactor = 7.3 m^2)

Figure 6.12 shows the corresponding concentrations of selected electron acceptors (ClO_4^-, NO_3^-, and O_2) through the MBfR system. Through the system, influent nitrate and oxygen levels were reduced to below detection limits in the first reactor. Approximately 60 percent of the perchlorate reduction occurred in the first reactor and was further removed to 2 µg/L through the second reactor.

Dependence of Perchlorate Removal on Other Electron Acceptors

The results of Phase I of this project, as well as the results of other perchlorate biodegradation studies, have demonstrated that perchlorate biodegradation is slowed by elevated nitrate concentrations. However, there is little information on whether perchlorate biodegradation

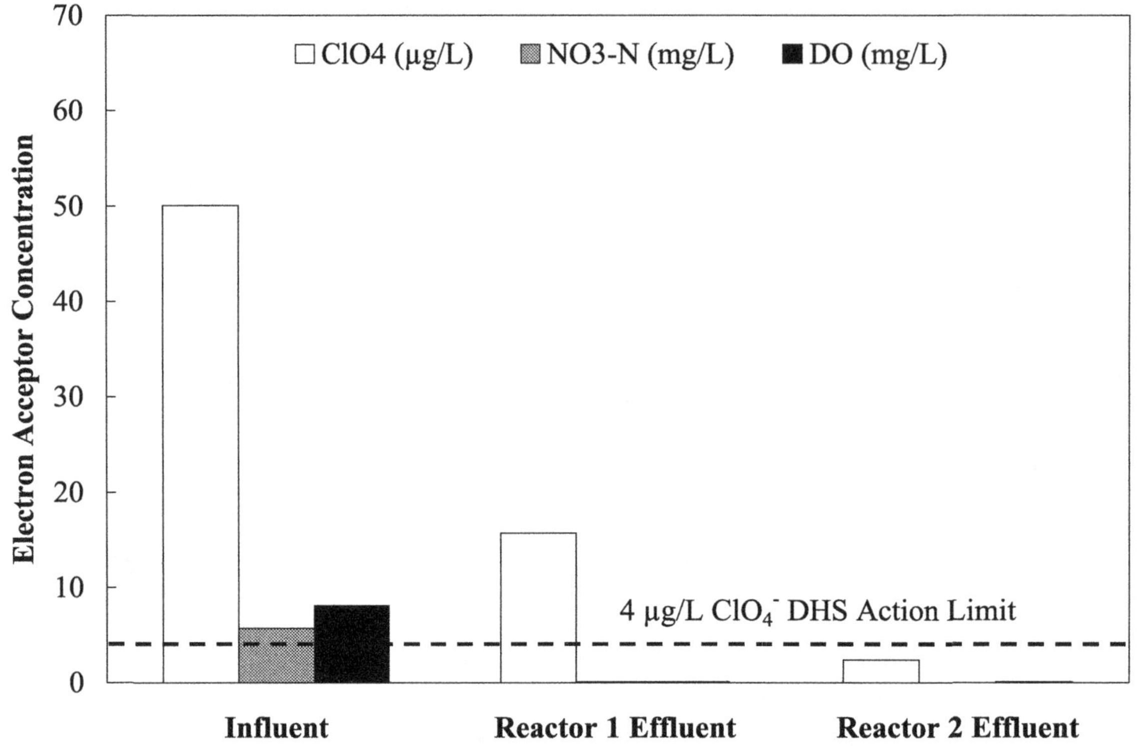

Figure 6.12 Reduction of oxygen, nitrate, and perchlorate through the MBfR (Q=0.15, R=27)

can take place at all in the absence of nitrate in the feed water. This is a critical question to answer because many perchlorate-contaminated groundwaters across the US may not contain nitrate. Therefore, it is essential to determine whether this process is limited to waters containing nitrate, or is applicable to all perchlorate-contaminated groundwaters. The concern is that alone, low perchlorate concentrations may not serve as a sufficient energy source for biological growth to occur. Consequently, the presence of nitrate may be necessary for this process to sustain a biofilm.

Nitrate

During steady-state operation after the reactor acid cleaning, influent nitrate (5.6 mg-N/L) and dissolved oxygen (8 mg/L) were reduced to below detection limits in the first reactor. Additionally, the influent perchlorate (55 μg/L) was reduced by approximately 60 percent to 20 μg/L. As a result, it was possible to evaluate the dependence of perchlorate removal on nitrate concentration in the second reactor. As shown in Figure 6.13, once steady-state removal of perchlorate (in the absence of nitrate) had been established, the following three conditions were evaluated for perchlorate removal:

- Transient spike of nitrate
- Continuous ambient groundwater nitrate concentrations
- Absence of nitrate, but following the development of an actively denitrifying biofilm

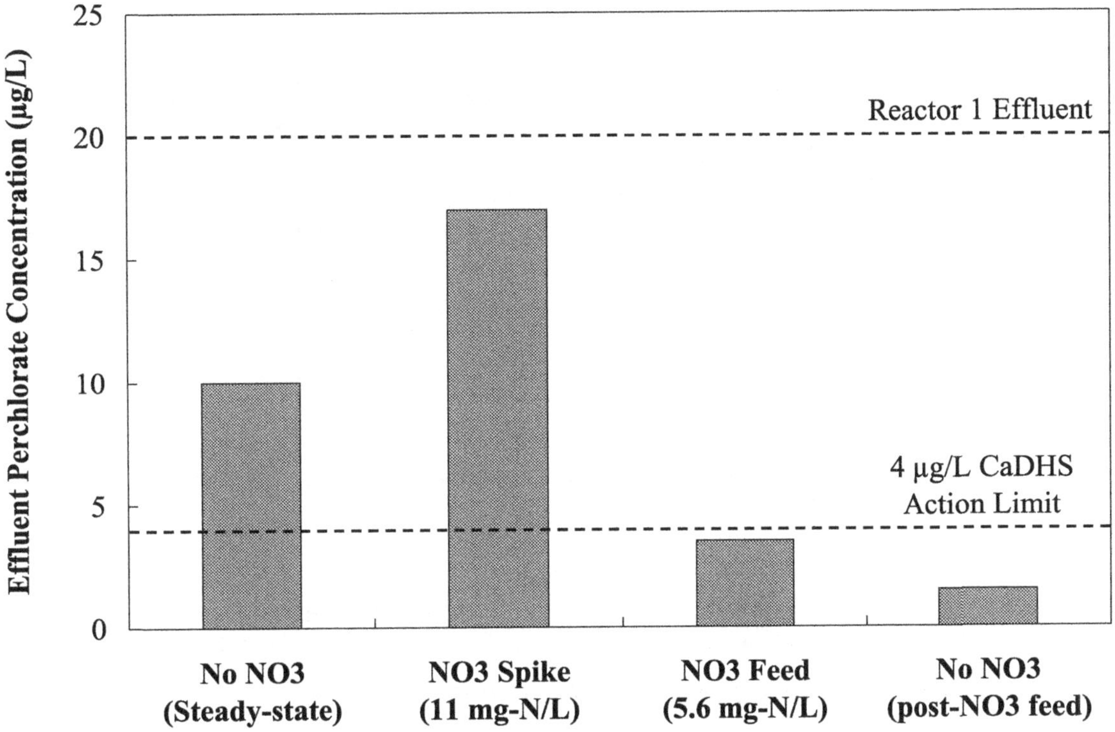

Figure 6.13 Dependence of perchlorate removal on nitrate concentration

Initially the second reactor was removing 50 percent of the perchlorate, reducing its concentration from 20 to 10 µg/L. While continuous performance indicated that this appeared to be at steady state conditions, it was not possible to determine if this amount was the result of the mixed perchlorate reducing biofilm culture being solely sustained on the reduction of 10 µg/L perchlorate, if the PCRB were sustained on oxygen or nitrate, if this was could be attributed to a continuous seed of actively perchlorate reducing organisms from the first reactor, or a combination of two or more of these possibilities.

Once a stable perchlorate concentration (10 µg/L) was observed out of the second reactor for a 2-week period, the first test condition was evaluated. A 1-day transient 11 mg-N/L nitrate spike, double the ambient groundwater concentration, was added to the influent of the second reactor. During this spike perchlorate reduction deteriorated from 50 to 15 percent and approximately 1 mg-N/L nitrate was observed in the reactor effluent. A corresponding drop in the dissolved hydrogen concentration to below detection indicated that the system was electron donor limited. While this finding was not unexpected, it demonstrated that in the event that hydrogen is limited, nitrate will be preferentially reduced, but some perchlorate will still be reduced. It is hypothesized that this perchlorate reduction will occur in the inner most portion of the biofilm where the hydrogen concentration is highest and the nitrate concentration is lowest.

Once the second reactor influent nitrate concentration was reduced to 5.6 mg-N/L, complete denitrification was observed, and perchlorate reduction improved, quickly reducing the effluent perchlorate concentration to 3.5 µg/L. Once the 7-day nitrate feed was terminated in the second reactor, perchlorate reduction further improved, reducing the effluent perchlorate

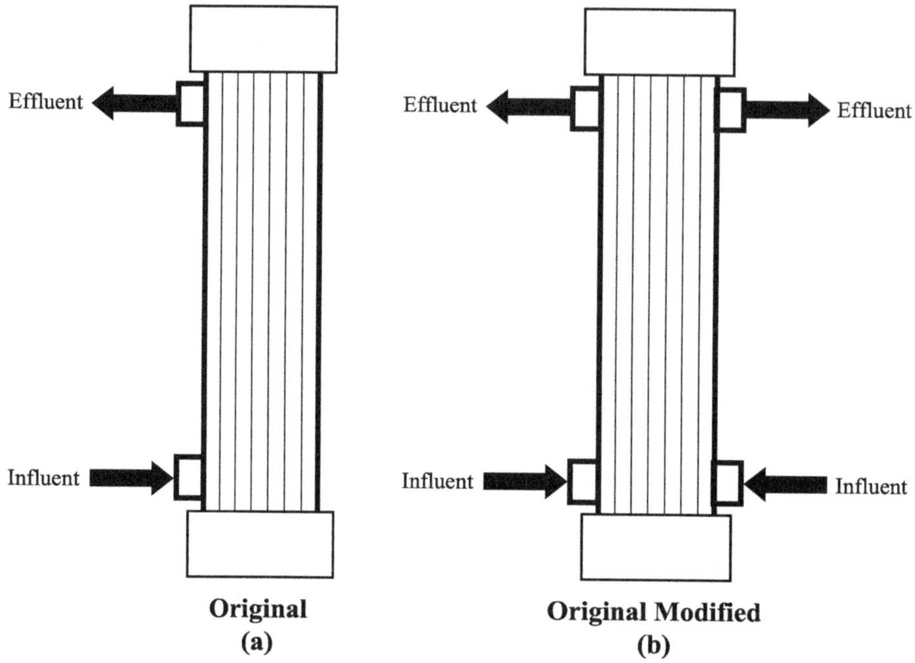

Figure 6.14 Original (M3) versus modified (M3) module designs

concentration to 1.5 μg/L. This performance, expected as the increased biomass adapted to nitrate and perchlorate, was acceptor-limited and consumed more perchlorate in the absence of nitrate.

OPERATION AND MAINTENANCE ISSUES

During pilot plant operation, various operational and maintenance issues were identified as critical to implementation and successful performance of a hydrogen-fed hollow-fiber membrane biofilm reactor. These issues focused on:

- Recycle Flow
- Reactor Cleaning
- Fiber Integrity
- Reducing Complexity

Recycle Flow

The design of the initial reactors was based on conventional full-scale, low-pressure microfiltration/ultrafiltration membrane modules. Thus, the reactors have a single influent and a single effluent port for the liquid on the outside of the membrane fibers, perpendicular to the length of the reactor. These ports are located in-line with each other on the same side, and they are close to the ends of the module (Figure 6.14a). While this configuration may be efficient when filtering water, this design provides poor distribution of flow when the water passes along the outside of the membrane fibers. To overcome the poor flow distribution, we initially used a high recycle flow rate (R=50). The high recycle flow reduces short-circuiting and minimizes mass transport resistance. However, the high recycle flow rate incurs significant pumping costs and excessive headloss through the reactor.

Preliminary work was done to modify these original reactors to evaluate if hydraulics could be improved and if the high recycle flow rate could be lowered. The modification of the original design entailed adding a second influent and effluent port axially, across from the existing influent and effluent ports. Using this configuration (Figure 6.14b), influent water entered from the bottom of the reactor opposite from each other, traveled up the reactor and exited in a similar fashion. An immediate improvement with the original modified reactors was observed as the water was more evenly distributed through the reactor. With the original reactor design, perchlorate removal below 4 μg/L was achieved at a system flow rate of 0.15 gpm (HRT = 56 min), but a recycle flow rate of 7.5 gpm (R=50) was also necessary. With the modified reactors, the same removal was being achieved at nearly half the recycle flow (R=27). No further reduction in the recycle ratios with these modified reactors was evaluated, as the reactors needed to be redesigned. This preliminary reactor modification of the two-reactor-in-series system demonstrated that a minor improvement of the system configuration could produce, significant hydraulic improvements (R=50 to 27). Consequently, we focused our efforts on the new single-reactor designs and not on optimizing a system that would not be implemented.

Reactor Cleaning

During long-term operation, the reactors must be cleaned to maintain their effectiveness. Without cleaning, several problems may develop:

- *Excess Biomass.* The build-up of excess biomass can lead to the clumping of fibers, which reduces the effective biofilm area, exacerbates short-circuiting, reduces mass transport of perchlorate to organisms in the biofilm, and increases the distance hydrogen must diffuse to reach the outer portion of the biofilm. Each of these can result in decreased performance of the system.
- *Sulfate Reduction.* If the biomass is not regularly exposed to oxygen and excess biomass is not frequently removed (i.e., slow growing organisms are allowed to proliferate), sulfate-reducing organisms may establish themselves in the biofilm. Once established, the reduction of sulfate and concomitant production of hydrogen sulfide will result in a mild aesthetic problem. However, the subsequent aeration and chlorination processes oxidize any residual concentration of sulfide to sulfate.
- *Calcification.* The raw water entering the pilot plant was supersaturated with respect to calcium carbonate (Langelier Saturation Index = 0.7 pH units, Calcium Carbonate Precipitation Potential = 21 mg/L as $CaCO_3$). The calcium carbonate precipitation potential (CCPP) increases with the resulting increase in pH observed with denitrification. Over time, calcium carbonate can precipitate on the walls of the reactor and deposit within the biofilm. If the calcium carbonate in the biofilm is not regularly displaced, it can accumulate, leading to the calcification of the fibers. Calcification of the fibers reduces their flexibility and can result in fiber breakage when the hydraulics through the reactor change with varying flowrate or air-scour. It can also result in decreased hydrogen transfer if the calcium precipitates within the pores of the fiber.

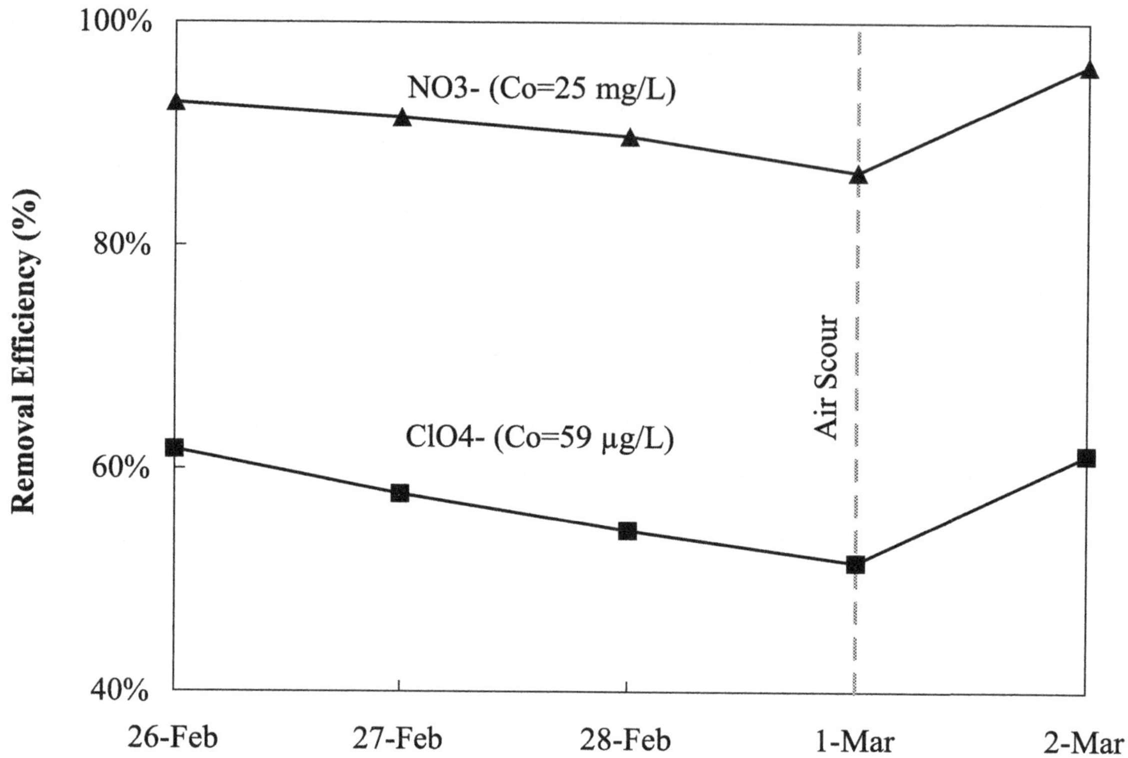

Figure 6.15 Effect of air scour on MBfR system performance

Raw water calcium was regularly measured at 62 mg/L as Ca^{2+}. At steady-state conditions, approximately 30 mg/L of calcium was precipitated through the reactors. Further reduction of calcium through the aeration basin and media filter was not detected. Over time, the areas within the reactors that were not well cleaned developed a thick white chalky colored overgrowth. Samples of this overgrowth revealed severe calcification with a calcium content of 105,000 mg/kg. This indicated that over 25 percent of this overgrowth was calcium carbonate.

Air Scour

On several occasions, air scouring of the membrane fibers was required to prevent and/or remove excessive biomass buildup. If unchecked, the excessive biomass resulted in the decreased performance of the system. Clumping of fibers, decreased dispersion of the flow throughout the reactor, and increased distances that the electron acceptors must diffuse from the bulk flow to the dense inner biofilm are some of the likely reasons for the observed decrease in the performance of each reactor.

To air scour a module, a jet of compressed air was released into the influent located at the bottom of the module. While under otherwise normal operating conditions, the air traveled up the reactor vigorously agitating the membrane fibers, shearing loosely attached biomass, and leaving only the dense biofilm around the individual fibers.

Initially during operation, an air scour was initiated when individual membrane fibers were not visually discernable from each other, resulting from the clumping of several membrane

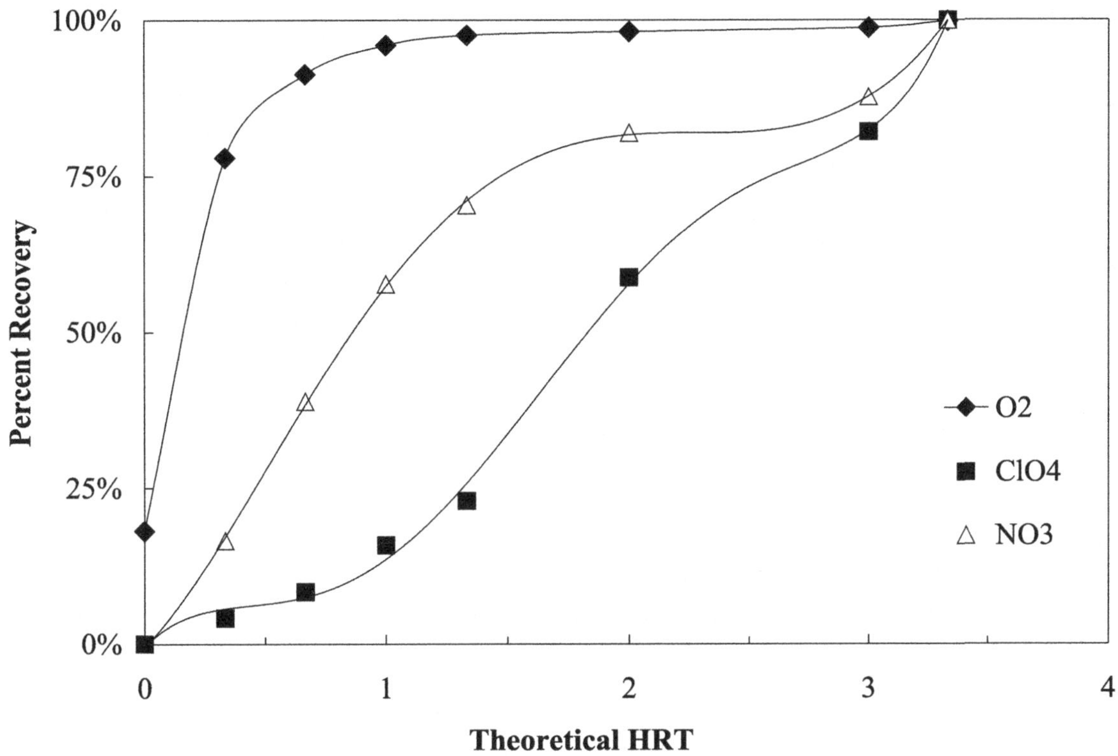

Figure 6.16 Reactor recovery after air scour cleaning

fibers into large bundles, or a decrease in the MBfR system performance was detected. Figure 6.15 shows the decreased performance of the MBfR system as an excessive amount of biomass was allowed to develop, and the resulting improvement after an air scour. Each air scouring had demonstrated a complete recovery of the performance decrease attributed to excessive biofilm development. While operating at a flow rate of 0.3 gpm and a recycle ratio of 15, air scouring typically recovered all of the 3 to 10 percent lost of overall reactor performance.

If the biomass was allowed to buildup until the majority of the fibers had clumped together, a conventional air scouring was not sufficient to effectively separate the fibers. An alternative module design would be likely to improve the efficiency of the current air scouring process. With the existing design, however, a "physical" scouring was required to separate the fibers. The physical scouring involved dismounting individual modules of the MBfR from the system panel and shaking the module only partially filled with water until the individual membrane fibers were separated. This method was more effective at breaking up clumped biomass, especially near ends of the module, as compared to conventional air scouring during normal operation. However, additional effort was required to perform this physical scouring of the membrane fibers and the process was effectively off-line for longer periods.

Despite the system interruption that resulted from either an air or physical scouring, the pilot system quickly recovered and returned to normal operation and performance within three theoretical hydraulic retention times (Figure 6.16). This observation demonstrated that the indigenous bacteria of the dense biofilm were resilient and robust, and the bacteria sloughed off during scouring were not necessary for effective perchlorate, nitrate, and oxygen reduction.

Acid Cleaning

While the air scouring was an effective regular cleaning process, periodic cleanings that are more rigorous in nature may still be required to regulate calcification in the "difficult to air scour" areas. This rigorous cleaning process was developed after consulting the membrane manufacturer for the fiber chemical resistance data and recommended cleaning procedure. Based on the information provided, a 3 percent citric acid solution was used to clean the membranes. The membranes were soaked in this solution for one hour with recycle, to ensure that all regions of the reactor were uniformly treated, and then rinsed with clean water before returning to service.

Through this acid cleaning, the majority of the precipitated calcium was visually removed. Additionally, the hydrogen concentration was measured at nearly saturation (2000 μg/L) as compared to the 50 μg/L measured at the same operating gas pressure immediately before the acid cleaning. Interestingly and importantly, the fibers still maintained a visible and active biofilm. Samples collected 24 hours after the cleaning revealed an improved reactor performance with respect to all of the electron acceptors. This showed that a mild acid clean to remove calcification was possible without harming the active biofilm.

Given the success of the initial acid clean, a more aggressive acid clean was performed to determine if it could be used to completely remove all visible signs of calcification without harming system performance. The aggressive acid clean of the first reactor in series included an initial one-hour soak with recycle followed by an extended soak with fresh citric acid for six-hours. After the six-hour soak, the precipitated calcium was completely dissolved; however, it visually appeared that the biofilm had been completely removed, exposing the membrane fibers in a significant portion of the reactor.

The system was placed back into service, at a flow rate of 0.20 gpm (45 min HRT) and a recycle flow of 5 gpm (R=25), and monitored closely to evaluate system recovery after the aggressive acid clean. Within a 72-hour period, greater than 95 percent removals of both dissolved oxygen and nitrate were measured. As shown in Figure 6.17, perchlorate removal in the first reactor was minimal for the first four days and then improved linearly over the next couple weeks, after which perchlorate removal through the first reactor returned to within normal removal rates (50–60 percent). This shows that with a minimal seed (the residual sparse biofilm remaining after the aggressive acid clean), the process can be quickly started up.

Fiber Integrity and Repair

The first fiber breakage was detected after approximately four months of operation. The number of compromised fibers dramatically increased after the original reactors were modified to improve distribution of the influent water. Concerns of the reliability of the system were raised. While many hypotheses were presented as possible reasons for the increase in fiber failures, only two were seriously considered. These ideas included:

- Stretching the fiber past the point of elasticity. Because of the poor hydraulics through the reactors, high recycle flow rates were required to obtain the pilot-scale removals expected from the bench-scale results. At the influent and effluent orifice of the original reactors, the fluid velocity was 93 cm/sec (recycle flow of 7.5 gpm). The fibers close to this point of high velocity might be stretched over time to the point of breaking.

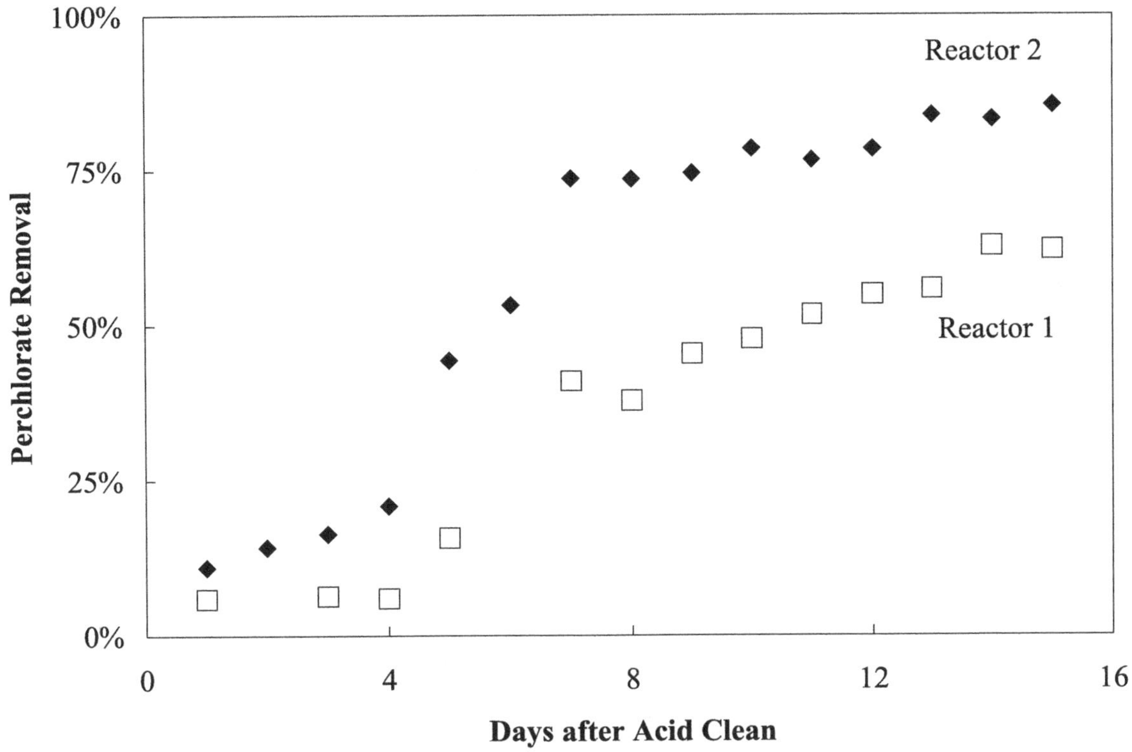

Figure 6.17 Reactor recovery after aggressive acid clean

- Calcification. The influent water to the pilot plant had a positive calcium carbonate precipitation potential of 21 mg/L as $CaCO_3$. Over time, the fibers that could not be effectively cleaned by air scour developed a thick white chalky colored overgrowth, and samples of this material revealed a calcium content of 105,000 mg/kg. Calcification could lead to a decreased flexibility of the fiber and, during periods of extreme turbulence (i.e., during air scour), the reduced flexibility might lead to fiber breakage.

Immediately after both reactors had been acid cleaned to remove the deposited calcium, the turbulence generated during air scouring no longer resulted in compromised fibers, as shown in Figure 6.18. If fibers were being stretched to the point of breaking by the high fluid velocities, it would be expected that even after the acid cleaning, fiber repair would be required. This was not the case and consequently, it was determined that calcification, not stretching of the fibers, was the primary cause of the fiber integrity problems. Additionally, periodic acid cleanings and regular air scour can prevent calcification leading to fiber breakage, even if the water used has the propensity to precipitate significant amounts of calcium carbonate.

Compromised fibers detected during pilot testing were repaired according to the following procedure. First all liquid connections to the MBfR module with the compromised fiber were disconnected and the entire module was removed from the system panel. The potted fiber ends were then exposed by removing the two module end caps used to contain the gaseous hydrogen. The module, now with end caps removed, was mounted back to the system panel and all liquid connections were reattached. To identify which fiber within the potting was compromised, the module was filled with water until a bead of water was observed coming out of each end of the

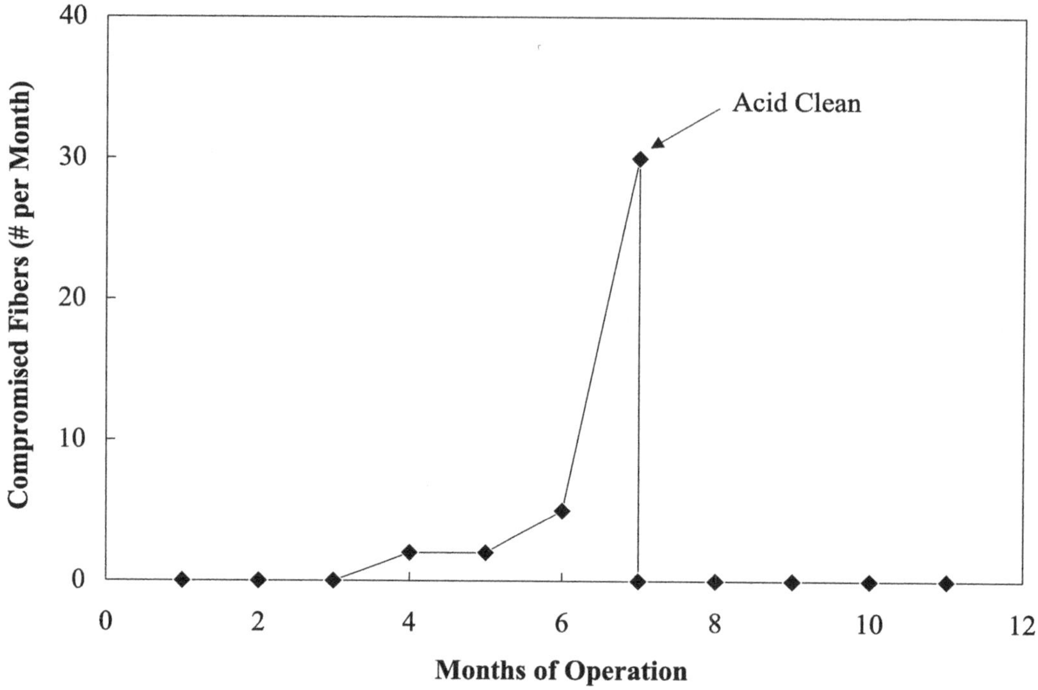

Figure 6.18 Evaluation of fiber integrity

Figure 6.19 Repair using soldering iron and membrane material

compromised fiber. The identified fiber of interest was then marked with a syringe needle for further repair.

To repair the identified fiber, a fine tipped soldering iron and extra membrane material was required. As outlined by the reactor manufacturer, the soldering iron was used to melt a 1-mm deep groove surrounding the compromised fiber (marked by the syringe needle) in its appropriate fiber bundle. A small amount of membrane fiber material was then melted to seal the end of the compromised fiber, as shown in Figure 6.19. The aforementioned groove was used to contain the melted membrane material and prevent it from spreading and blocking uncompromised fibers.

Once both ends of the compromised fiber were repaired, the module was pressurized with water and the repaired ends were checked for liquid leaks. The end caps were then reattached to the module and a gas leak test was performed on the module by pressurizing the end caps and lumens of the fibers with compressed air. If no air bubbles were observed, all liquid and gas connections were reconnected and the MBfR system was returned to service.

Reducing Complexity

The original system had two MBfRs in series. Each MBfR module had independent controls (process flow, recycle ratio, water pressure, and hydrogen pressure), as the requirements of each reactor were different. Additionally, the operation and control of the second reactor in series depended on the operation and performance of the first reactor. At full-scale, having multiple sets of two reactors in series would introduce a significant amount of complexity related to process control and increase the burden on the operators to understand and maintain the different biological mechanisms and processes. Therefore, a shift to a single-reactor system was proposed so that the complexity of the system could be significantly reduced. However, in order to achieve similar removal rates to that of the two-reactor-in-series system, the hydraulic limitations of the original modules needed to be addressed through reactor modification and redesign.

Process Interruptions

Several unplanned system interruptions occurred during the course of operation. Listed below are various scenarios observed during pilot testing and the length of time the MBfR system was offline. For each process interruption, the MBfR pilot performance immediately recovered to its prior state without diminished performance after fixing the source of the problem.

- System auto-shutdown due to influent magmeter failure (24hrs)
- Water source unavailable due to La Puente shutdown (120 hrs)
- System dewatered after tubing failure (24 hrs)
- Recycle pump failure (24 hrs)
- Fiber Compromised (24 hrs)
- Hydrogen feed gas failure (24 hrs)

As discovered while evaluating system recovery during air scouring, these favorable results indicated the presence of robust and resilient indigenous bacteria within the biofilm.

Preventative Maintenance

Based on these various findings, a preventative maintenance schedule was designed to:

- Control excessive biofilm growth
- Develop a dense resilient biofilm
- Avoid clumping of fibers
- Prevent sulfate reduction
- Minimize calcium accumulation on fibers

Air scour cleanings should be performed every other day at a minimum to control excessive biofilm growth and avoid clumping of fibers for the water quality tested. Air scour could be performed daily without negative impact on reactor performance. Regularly stripping the loosely held biofilm by air scour combined with high fluid velocities was effective in promoting the development of a dense robust biofilm. Sulfate reduction also was minimized through the regular exposure to oxygen introduced during air scouring. Finally, calcium deposition primarily occured on the external portion of the biofilm and was removed from the system during air scour.

Prior to the aggressive acid cleaning, air scouring was only performed once every two weeks. Significant calcification was noticeable after approximately two months. Since the acid cleaning of both reactors and the implementation of air scouring every other day, no visible calcification of the fibers was observed over a four-month period. Solids, however, accumulated in the bottom of the reactors in the areas that the air scour did not effectively agitate. Consequently, a monthly mild acid clean was performed to remove calcium that accumulated with these solids. If solids accumulation could be reduced or completely avoided through improved hydraulics and more efficient air scour cleaning, mild acid cleaning on an semi-annual basis or even less frequently may be sufficient to protect the system from calcification and ultimately, loss of fiber integrity.

CHAPTER 7
REACTOR OPTIMIZATION

Although both the original MBfRs successfully reduced the perchlorate concentration to below the 4-µg/L CaDHS Action Limit, the economics and reliability of the system needed to be significantly improved by addressing shortcomings of the current configuration. Three modified and redesigned MBfRs were evaluated to address the design related operation and maintenance issues. These included:

- *High Recycle.* The original configuration required unusually high recycle flows to account for the poor flow distribution through the reactor. Redesign necessitated an alternative method to ensure an even distribution of flow through the fiber mass to avoid uneconomical high recycle flow rates.
- *Air Scour.* Although the original design allowed for air scour, it did not efficiently clean all of the fibers. Redesign with an alternative introduction of air and/or water for an improved pneumatic or hydraulic cleaning was important for regular maintenance.
- *Acid Cleaning.* Periodic acid cleaning of the fibers to remove calcium deposits was identified as a necessary part of the maintenance of the system to ensure fiber integrity. Any new design must not be impacted by prolonged exposure to acidic conditions (3 percent citric acid).
- *Complexity.* The original configuration had two reactors in series, with each reactor requiring a separate set of operational conditions and control set points. At full-scale, having multiple sets of these reactors in series would introduce a significant amount of complexity related to process control and increase the burden on the operators to understand and maintain the different biological mechanisms and processes. The new systems will be designed to operate as a single-reactor system to significantly reduce the complexity of the system. This change may also reduce the cost for construction.

The two manufactures selected for module redesign efforts were Mitsubishi Rayon, Corp. (MRC) and Membrana (previously Celgard). Our discussions with the manufacturers resulted in different approaches to address the various design issues. Additionally, different hollow fiber membranes were used. MRC uses a composite fiber consisting of three layers; a dense polyurethane layer sandwiched between two hydrophobic microporous polyethylene layers. Membrana uses single-layer microporous polypropylene membrane fibers. A summary of the physical characteristics of the systems discussed to date is provided below.

The following sections document the performance of each single reactor system tested and include:

- Membrana Liqui-Cel
- Modified Mitsubishi - Looped Fiber Design with 20 Percent Packing Density (M20)
- Original Design Mitsubishi - 6 Percent Packing Density (M6)

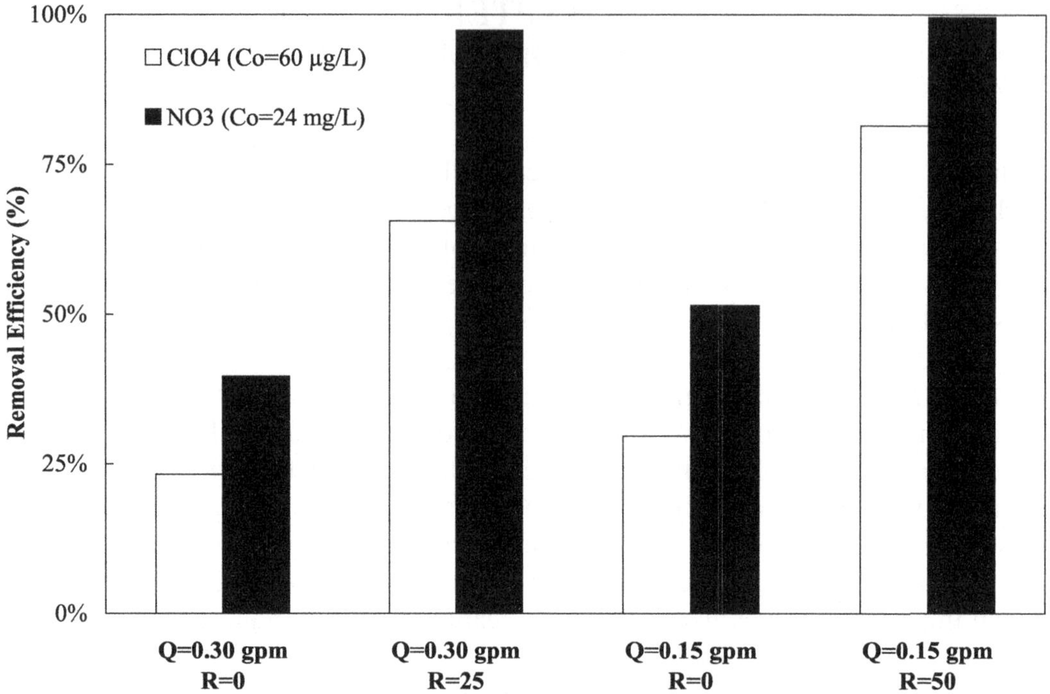

Figure 7.1 Influence of recycle on fractional removal of nitrate and perchlorate

HYDRAULIC LIMITATION OF THE ORIGINAL DESIGN

Additional hydraulic tests were performed with the original reactor design at flowrates of 0.15 and 0.30 gpm to further demonstrate the hydraulic limitations and the effect recycle flow had on system performance (Figure 7.1). For each condition, separately recycling the water through each reactor at 7.5 gpm (R=25) resulted in an approximate tripling of the percent perchlorate removal and a doubling the nitrate percent removal.

Interestingly, comparing the same data based on the total mass removed or apparent substrate flux revealed that the mass flux of perchlorate or nitrate reduced was only 1.5 times higher at 0.30 gpm as compared to 0.15 gpm, as shown in Figure 7.2. This demonstrated some of the bioreduction kinetic limitations of the system, as the flux did not double with a doubling of the flow rate. The dramatic improvement with recycle, however, revealed that serious hydraulic limitations (substrate gradients) exist for the current module design without recycle.

Also, it is important to note that an immediate improvement in reactor performance was observed within one HRT after returning the system to service with recycle flow for both hydraulics tests.

MEMBRANA LIQUI-CEL

Membrana's Liqui-Cel reactor contained polypropylene hollow fibers separated using a knitted array that was tightly wound around a distribution and collection tube, as shown in Figure 7.3. Liquid entered the reactor and traveled through a perforated distribution tube, designed to improve flow distribution by forcing the water past all of the hollow fibers. A central

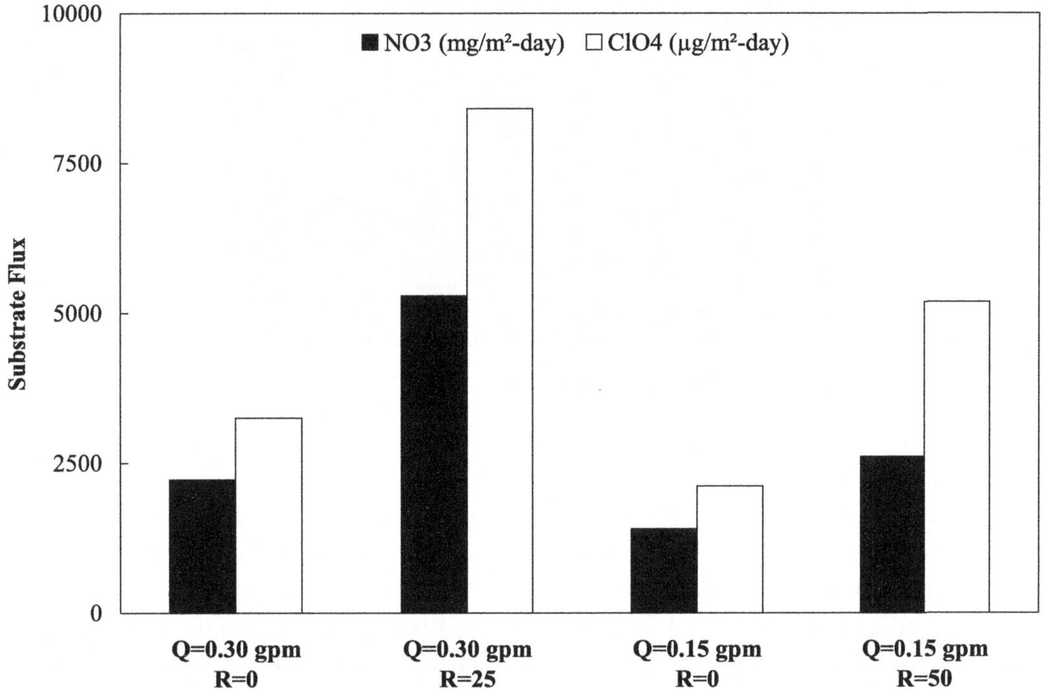

Figure 7.2 Influence of recycle on apparent substrate flux

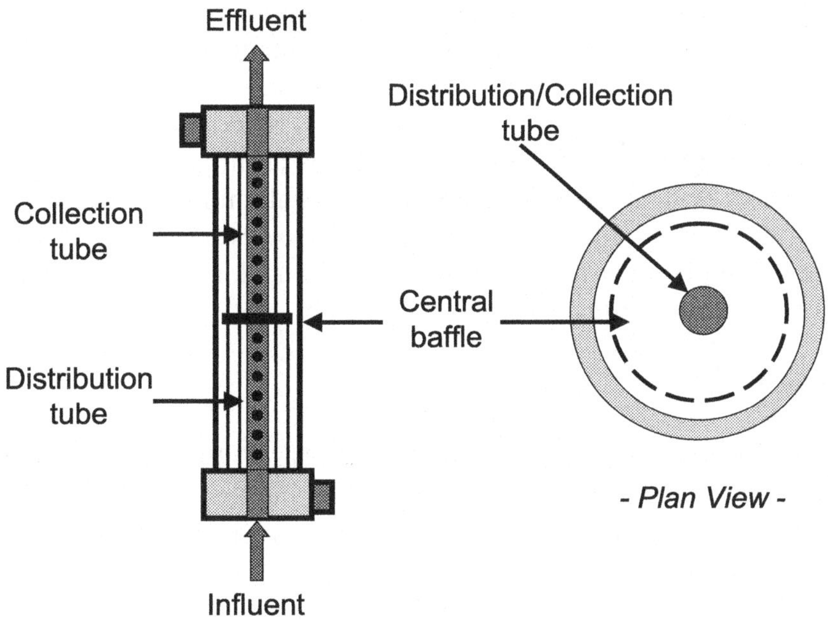

Figure 7.3 Membrana Liqui-Cel module

Figure 7.4 Membrana hollow fiber knitted array

baffle located midway through reactor, directed the liquid around to the second half of the reactor, at which point the liquid again travelled through the fiber bundle and entered the collection tube, perforated similarly to the distribution tube. These unique features (perforated distribution and collection tube, center baffle) ensured that the water would come into contact with all of the fibers regardless of the flowrate. As with the rest of the MBfRs, hydrogen gas was fed to the upper and lower ends of the reactor to diffuse through the lumens to the biofilm growing on the outside of the fibers.

Figure 7.4 shows the knitted array that ensures all of the fibers are held an equal distance apart from each other (~100 μm). In contrast to the composite hollow fiber membranes, the membrane used in this reactor has only a single microporous layer. The single-layer hollow fiber functions identically to the multilayer composite fibers previously used in Mitsubishi modules. The main difference between both types of membranes is that the composite fibers are physically capable of resisting bubbling at higher gas pressures. For conventional applications of these hollow fibers membranes (aerating or de-aerating), this is an important feature, as the higher gas mass transfer is possible with composite fibers while maintaining the integrity of the membrane fiber and extending the bubble point of the membrane. However, for the MBfR application, this feature (composite vs. single-layer microporous) is minimized, as the hydraulic and gas pressures are controlled to be maintained equal to each other.

System Performance

The Liqui-Cel reactor was operated with an HRT of 30 minutes and a recycle flow of 2.2 L/min (Recycle Ratio = 10, linear velocity = 17 cm/min). The reactor was seeded with biomass developed from previous operation of the original Mitsubishi MBfR (3%) and reached steady-sate operation within one week.

For the first three weeks of operation (Figure 7.5), influent DO and nitrate levels were reduced to below detection limits. Perchlorate reduction, however, was limited to approximately 20 percent. During the second week of operation, recycle flow was terminated due to excessive headloss experienced through the reactor, as high as 50 psi. It was impractical to operate the system long-term under such a high pressure, as the system was not designed for it. Additionally, the hydrogen pressure was kept at the low effluent pressure to ensure that the bubble-point was not exceeded at any point along the length of the module. Once the recycle flow was eliminated,

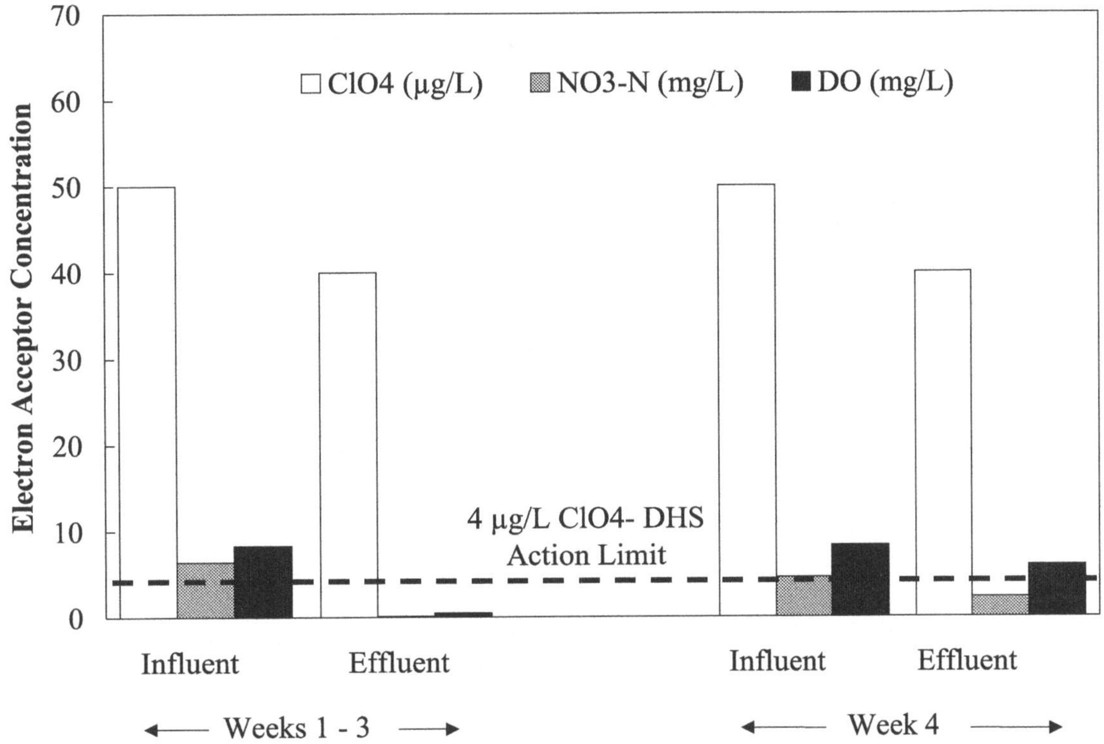

Figure 7.5 Summary of Liqui-Cel performance

interestingly, no change in the steady-state reduction of DO, nitrate, and perchlorate was observed. At this time, it appeared that the improved reactor hydraulics and mass transfer from the module design was able to offset the previous requirement of a moderate to high recycle flow.

Major limitations of the Liqui-Cel design were encountered during regular air scour cleaning, terminally affecting the performance of the system. First, the reactor had an extremely high packing density (27.3 percent) and fibers were potted in a tight spiral wound arrangement within the housing. Membrana's intended use of this reactor is for aeration and de-aeration purposes with a clean water. The dense knitted fiber array that promotes uniform hydraulics through the fiber bundle minimized movement of the fibers, dramatically reducing the ability of an air scour to slough off extra biomass. Second, the tight fiber mesh prohibited the effective removal of any biofilm that may be displaced from the fibers during air scour. Lastly, as the air for the air scour was introduced through the perforated distribution tube, the air did not flow evenly throughout the fiber mass further reducing its effectiveness.

The impact of ineffective cleaning became readily apparent during the fourth week of operation. During this period, the reactor performance significantly suffered as the biofilm increased, resulting in hydraulic short-circuiting and the concurrent breakthrough of DO and nitrate (Figure 7.5). After continuous air scouring and even mild acid cleaning, system performance was unable to be restored, and the Liqui-Cel reactor was taken off-line.

Liqui-Cel Module Investigation

As a means to understand the cause of the Liqui-Cel's failure, a thorough investigation was performed by physically dissecting the module. A small longitudinal slice of the reactor

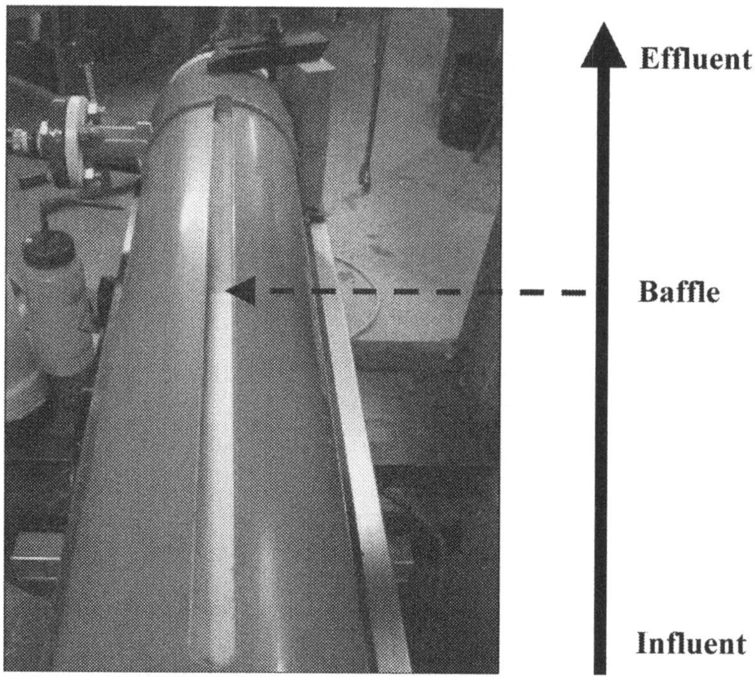

Figure 7.6 Liqui-Cel membrane fibers revealed

housing was removed to evaluate the initial status of the membranes. Interestingly, seen in Figure 7.6, there was a large visual difference between the two halves of the membrane cartridge. The lower half of the membrane cartridge (influent side) up to the baffle was fairly clean and free of biomass. However, the upper half of the membrane cartridge (effluent side) was covered with obvious biomass growth.

These observations of the outer fiber areas led the project team to hypothesize that the membrane fouling by biomass overgrowth was occurring primarily in areas in which the water initially entered the membrane fiber mesh. As evident above for the upper half of the module, the biomass accumulated on the outer portion of the fiber bundle. Conversely, the majority of growth accumulated next to the perforations in the distribution tube on the lower half of the module, where the raw water entered the fiber bundle through the distribution tube.

Liqui-Cel Conclusions

The main focus of testing the Liqui-Cel design was to compare performance and design features to that of the MBfR modules previously tested. The following is a summary of key design elements of the Liqui-Cel design.

- Single layer microporous hollow fibers
- Polypropylene fibers with hydrophobic characteristics
- Patented flow distribution system with the module to improve hydraulics
- High membrane surface area

All of these features are highly desired and optimal for gas transfer applications. However, some aspects of these design features were detrimental to the operation and performance of the module for its application as an MBfR. Based on the evaluation of the Liqui-Cel design operated as an MBfR, the project team was able to identify additional key features, outlined below, that would be required for successful operation and design of an optimized MBfR module.

- *Reactor Cleaning/Solid Removal.* While the importance of reactor cleaning was documented during operation of the initial MBfR system, its critical nature was identified here. As the biofilm was not effectively displaced, solids accumulated within the Liqui-Cel reactor, resulting in increasing headloss through the system and serious short-circuiting. This quickly led to a reduction and ultimately a failure of the system's ability to remove dissolved oxygen, nitrate, and perchlorate. Consequently, increased fiber packing density and improved hydraulics alone cannot supplant effective fiber cleaning and solids removal.
- *Membrane Fiber Packing Density/Surface Area.* The Liqui-Cel reactor contains more than 60,000 hollow fiber membranes, resulting in a fiber packing density of 27 percent and a total membrane surface area of 42 m^2. The M3 reactor contained only 8,000 fibers (membrane surface area = 7.7 m^2) and had a packing density of only 3 percent. Failed operation of the Liqui-Cel reactor was attributed, impart to the extremely high packing density and inability for fibers to freely move during air scouring. Although a high surface area may allow for more reduction to take place and reduce or eliminate the need for recirculation, it was obvious that a 27 percent packing density was too high. Additional testing must be done to evaluate the optimal membrane packing density, while maximizing membrane surface area and ensuring that effective cleaning of the fibers if possible with regular air scour.
- *Membrane Fiber Material/Composition.* Evaluating different types of hollow membrane fibers from different manufacturers was of interest to the project team. The Liqui-Cel reactor uses Membrana's polypropylene single-layer microporous membranes. In previous testing, only Mitsubishi Rayon's composite hollow fiber membranes were tested. Mitsubishi's composite fibers, composed of nonporous polyurethane layer sandwiched between two microporous polyethylene layers, were designed to withstand higher gas pressures and to reduce the potential for bubbling of hydrogen through the microporous layer controlled by the middle layer. For operation in a gassing/degassing application, this feature is important because the shell-side (water) and lumen-side (gas) pressures may not always be in balance with each other. However, from performance and safety points of view, it is important that the lumenside pressure of H_2 be controlled independently of the shellside water pressure while avoiding the bubbling. In this regard, composite membranes have a significant advantage over single-layer membranes, as the hydrogen pressure can be adjusted without exceeding the bubble point of the membrane.

MODIFIED MITSUBISHI—LOOPED-FIBER DESIGN WITH 20 PERCENT PACKING DENSITY (M20)

Modifications of the original Mitsubishi MBfR design were performed by Mitsubishi Rayon Corporation (Japan) and shipped to the project team for evaluation. Figure 7.7 presents a

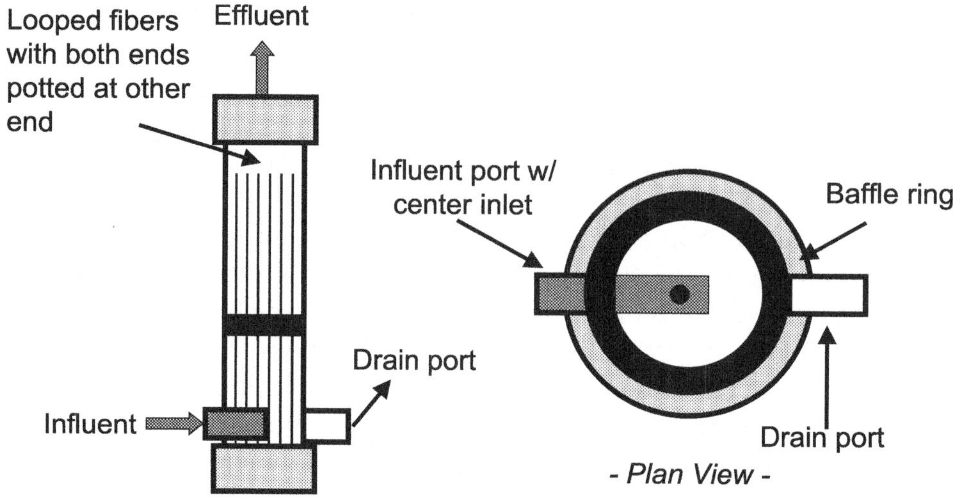

Figure 7.7 M20 module

conceptual diagram of this modified module (M20), which incorporated several different design features to improve performance. The following details key feature of the M20 design.

- *Hydraulics.* The M20 design included three new features to the MBfR module designed to improve hydraulic performance. To address the project team's concern about short-circuiting with the MF/UF designed reactor having influent and effluent ports on the same side of the reactor, the manner in which water entered the reactor was modified. Included in this design is a ¾-inch I.D. influent port with a small ¾-inch center inlet located in the center of the reactor. The center inlet was oriented so that water entered the reactor and was directed downwards toward the fiber potting. Mitsubishi's reasoning was that the entry of water in this manner would create a more even flow distribution as it moves up the reactor in tandem with the second feature. Approximately a third of the way downstream from the influent port was a central baffle ring. Approximately one inch in thickness, the central baffle ring was designed to redirect the flow along the sides of the reactor through the fiber bundle. Finally, the outlet port was modified such that water exited the reactor through the center of the end cap. This outlet feature was made possible by an alternative membrane fiber arrangement and potting.
- *Membrane Fiber Potting.* The same composite hollow fiber membranes were used in this new module. However, instead of fibers potted at both end of the reactor, they were looped in half, with both ends of the fiber potted at the influent end of the module. The looped "end," technically the middle of the fiber length, was free to move about at the effluent end of the module. This looped-fiber free-end design may be advantageous, because the fiber movement is not restricted as they are when potted at both ends. Having nearly a full range of motion allows air scour cleaning to more effectively jostle fibers to remove biomass solids that accumulate. In addition, it makes it possible for water to exit the reactor from the center of the end-cap rather than the side of the reactor. The drawback from this design is that there is no quick way to evacuate any condensation that may collect in the fibers, thereby decreasing hydrogen transfer.

- *Membrane Surface Area.* Mitsubishi also increased the packing density from 3 to 20 percent (approximately 40 percent within the central baffle ring). Now with 50,000 fibers in nearly the same module housing, the membrane surface area was increased by more than 6 times from 7.7 m^2 (M3) to 48.4 m^2. This dramatic increase in membrane surface area was designed to ensure complete reduction of all electron acceptor species within in a single reactor configuration rather than two reactors in series. Mitsubishi also intended the increase in membrane surface area to allow for a higher loading of electron acceptors as well, which would increase system throughput.
- The modified Mitsubishi reactor was operated with HRTs between 10 and 30 minutes and recycle flows corresponding to linear velocities of 61 to 154 cm/min. The reactor was seeded with biomass developed from previous operation of the M3 design and reached steady-sate operation within one week. Throughout the course of testing, different operating conditions were applied to the system, and the system performance was evaluated with respect to:

  - Hydraulic Retention Time
  - Recycle Flow Rate / Linear Velocity
  - Effectiveness of Air Scouring Cleaning

Hydraulic Retention Time

The M20 module was initially operated with an HRT of 30 minutes and reached steady state within one week. Complete removal of nitrate and DO was observed, and 70 percent reduction of perchlorate was also observed.

Following operation of the M20 at an HRT of 30 minutes, the system flowrate was increased to evaluate performance at several different HRTs. As seen in Figure 7.8, complete reduction of DO was easily attained at even the shortest HRTs. Complete denitrification was maintained as long as the theoretical hydraulic retention time was greater than 15 minutes. At lower HRTs (less than 15 minutes), nitrate breakthrough occurred. The maximum perchlorate reduction was 70 percent for HRTs greater than 24 minutes. Influent perchlorate concentrations averaged 50 μg/L, corresponding to an effluent concentration of about 10 μg/L.

Figure 7.9 similarly summarizes reactor performance with respect to substrate flux. The apparent perchlorate flux declined dramatically when the HRT was less than 10 minutes. Nitrate breakthrough first occurred at an HRT of 10 minutes, which indicates that more nitrate entered the system than the biofilm could utilize. With excess nitrate present, it appears that some of the bacteria reducing perchlorate slowed that reaction, probably the result of inhibition. The results are consistent with the laboratory mechanisms results and results with the M3 reactor.

Effect of Recycle

One of the goals of the M20 reactor modification was to address the hydraulic limitation of the original design. In addition to testing the modified reactor with varying HRTs, the recycle flow also was altered, resulting in linear velocities ranging from 61 to 154 cm/min. The average linear velocity observed within the MBfR during all testing was 120 cm/min.

In the M3 design, reactor performance improved with increased recycle flow. Changes in recycle flow affect the local mass transport kinetics and the flow distribution through the module.

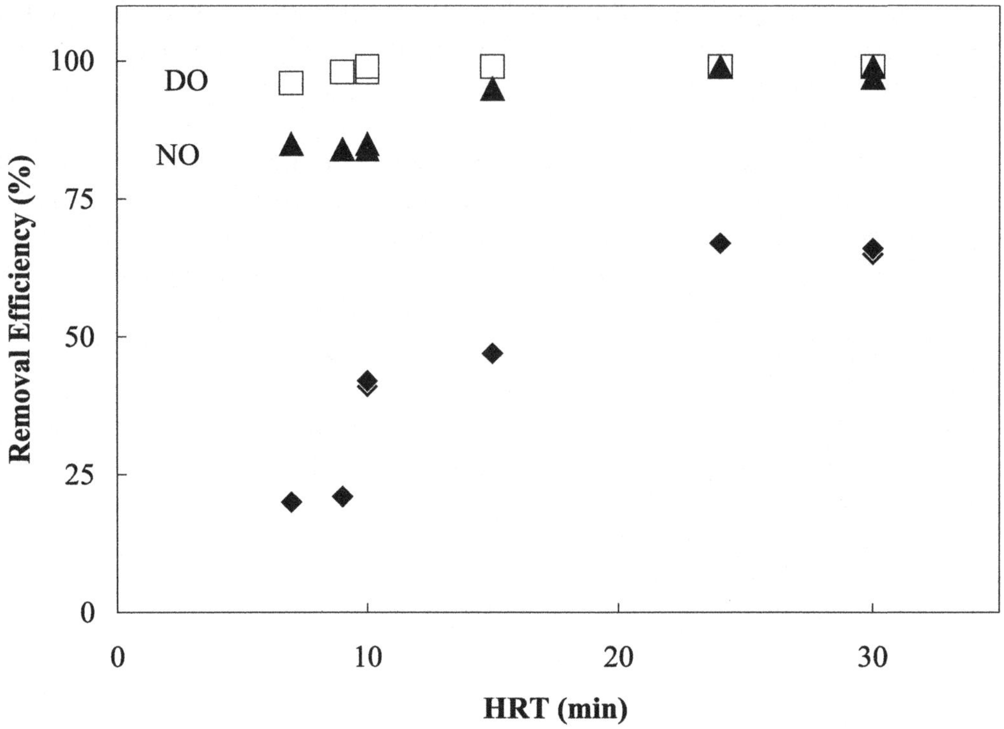

Figure 7.8 Influence of HRT on M20 electron acceptor removal efficiency

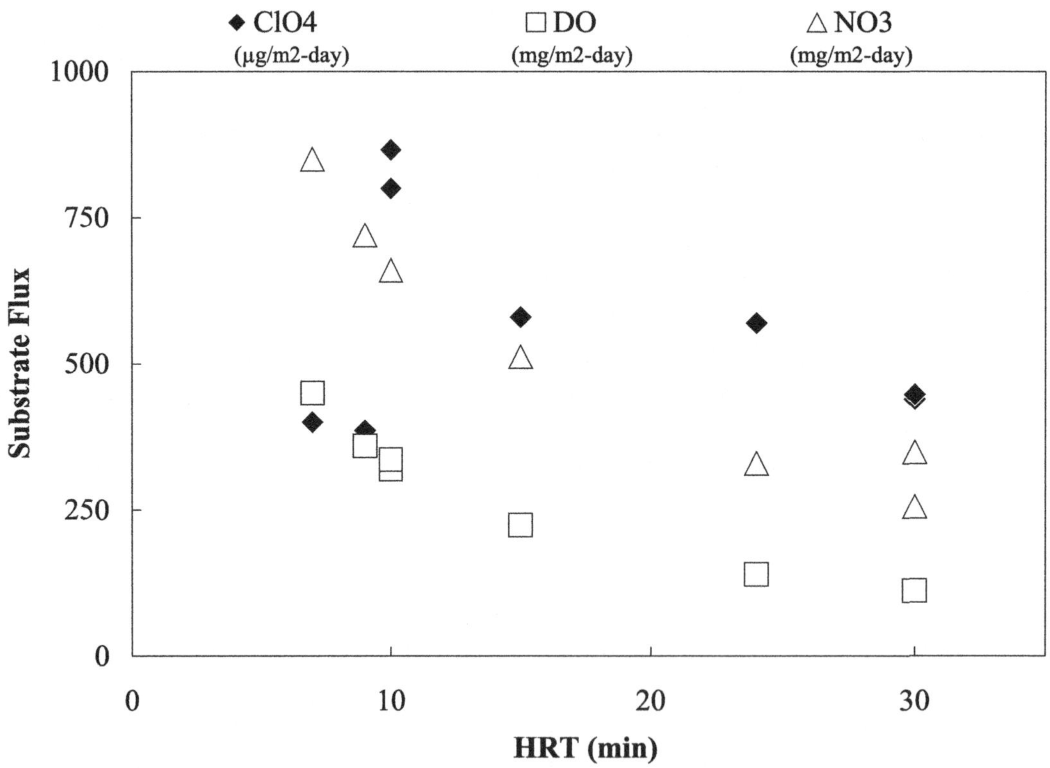

Figure 7.9 Influence of HRT on M20 substrate flux

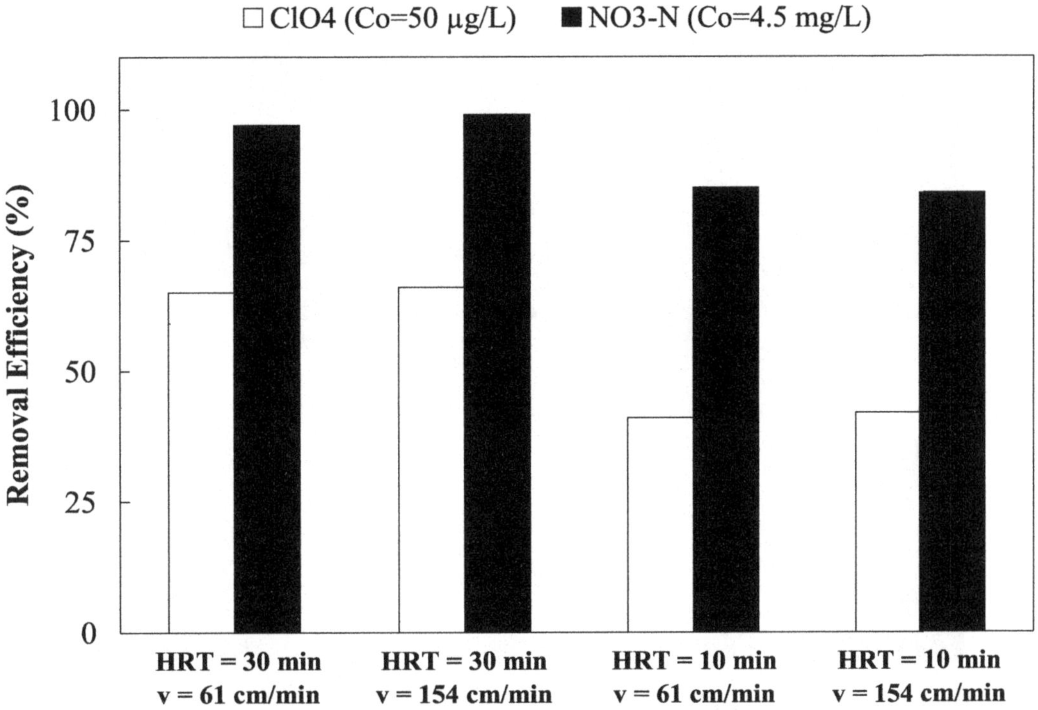

Figure 7.10 Effect of recycle on M20

The hydraulic shortcomings of the original design appear to have been improved with the M20 design. As summarized in Figure 7.10, the system was operated at two different HRTs, each at two different recycle flow rates corresponding to linear velocities of 61 and 154 cm/min. It is important to note that each condition represents steady-state operation, as the system was operated and monitored for a minimum of one week.

It can be clearly seen in Figure 7.10 that system performance did not improve when high recycle flow rates were applied for the same HRT. These results indicate that the M20 can be operated with a lower and more reasonable recycle flowrates while maintaining system performance. These data points favorably toward the economic feasibility of full-scale application of the MBfR by minimizing the capital and operating costs associated with providing a high recycle ratio.

Air Scour Modification

Reactor cleaning is critical to successful operation of the MBfR and is an especially important issue for the modified Mitsubishi reactor considering it contains 50,000 fibers (20 percent packing density). While the new features (baffle ring and influent ports) were improvements for hydraulics, they hindered effective air scour cleaning and solids removal with the high fiber packing density.

The M20 module contained a side drain port. The location was not ideal because it was located below the baffle. As fibers were concentrated within the baffle ring, the fiber packing density is increased from 20 to 40 percent at the ring. This made the movement of displaced solids past the baffle ring operationally difficult.

Figure 7.11 Biomass buildup before modified air scour

The anticipated advantage of having looped fibers potted at one end was freedom of movement of the fibers during operation and air scour. However, having the baffle ring located in the lower half of the module limited the movement of the fibers above the baffle ring during air scour. The fibers below the baffle ring also could not be effectively cleaned, as seen in Figure 7.11, because solids accumulated on the fibers and on the walls of the housing below the baffle ring. Overall, the baffle ring impaired hydraulics, cleaning efficiency was poor, and sulfate reduction occurred.

As a means to improve the method of air scour, the reactor was further modified by the project team by adding six air scour ports on the sides of the reactor housing, as illustrated in Figure 7.12. After adding the additional ports, air scour cleaning was improved and the biomass buildup below the baffle ring was minimized. In addition, air scour cleaning directly above the baffle ring was improved because an even distribution of air scour bubbles was allowed to travel up through the fibers.

M20 Conclusions

Mitsubishi Rayon, Corp. manufactured a modified version of the first generation MBfR by adding distinct features to address hydraulic limitations. The main focus of testing this design was to compare performance and design features to that of other MBfR modules previously tested. The following is a summary of key design elements of the M20 design.

- Looped fibers potted at a single end
- Baffle Ring for improved hydraulics
- Additional air scour ports added by the project team

Figure 7.12 Effect of air scour modification improvement

- High membrane surface area
- *Membrane Fibers.* Membrane fibers in the modified design are potted at one end of the reactor. This feature improved the fiber's freedom of movement and was designed to allow for a more effective air scour cleaning. The increased membrane surface area as compared to the original design resulted in similar reduction of DO, nitrate, and perchlorate to what was only previously achieved with the two-reactor-in-series M3 system. Perchlorate reduction, however, was limited to 70 percent.
- *Baffle Ring.* The baffle ring was intended to improve system hydraulics by limiting short-circuiting along the sides of the reactor and forcing water through the center of the fiber bundle. Although an improvement in the mass transfer, as lower recycle flows were sufficient to maintain steady-state reduction, the O&M associated with the location of the baffle posed problems. Cleaning efficiency was poor, as the packing density through the baffle ring was too high to allow sloughed biomass to pass. Reengineering may remedy this problem.
- *Air Scour Modification.* The need for improved air scouring was identified for this reactor design and was modified by the project team. The addition of several air scour ports was successful and would be recommended for future designs of this type. Not only does the inlet water need to be supplied to the module in an even distribution, but air scour must be supplied with the same principles in mind. The air scour bubbles tend to travel up the side of the reactor along the inside walls. Consequently, at least two strategically positioned air scour ports are recommended for even cleaning.

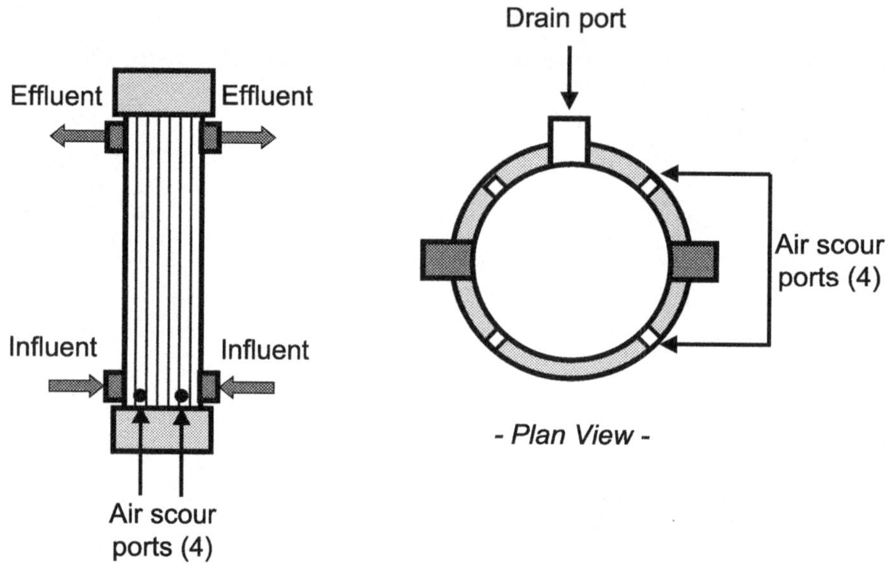

Figure 7.13 M6 module

ORIGINAL MITSUBISHI DESIGN—INCREASED PACKING DENSITY - 6% (M6)

Evaluation of the two alternative MBfR designs (Liqui-Cel and M20) highlighted points regarding membrane fiber packing density and air scour cleaning methods. Based on the understanding of the original design and the two alternative designs, a modified version of the original design was also created and tested. Figure 7.13 presents a conceptual diagram of this modified module (M6), which incorporates several different design features to improve performance.

- *Membrane Fiber Packing Density.* It was apparent from the testing of the Liqui-Cel and M20 reactors that 27 percent packing density was too high for successful application as an MBfR, and 20 percent may be too high if extreme steps for fouling control by proper air scour are not incorporated. The M3 design contained only a 3 percent packing density. For this modified system, we evaluated a module with a 6 percent packing.
- *Module Modifications.* The first modification of the M3 design included adding two influent and effluent ports so that water would be distributed in the reactor, reducing the potential of short circuiting. By adding additional influent and effluent ports, the existing original influent port could then be used as a drain port for solids removal. The second modification involved the addition of an improved air scouring system. Four air scour ports, independent of the influent ports, were added 90 degrees apart from each other to promote efficient and evenly distributed air scour.

The M6 reactor operated with HRTs between 10 and 30 minutes and recycle flows corresponding to linear velocities of 52 to 130 cm/min. The reactor was seeded with biomass developed from previous operation of the M3 reactor and reached steady-state operation within one week. Throughout the course of the testing, different

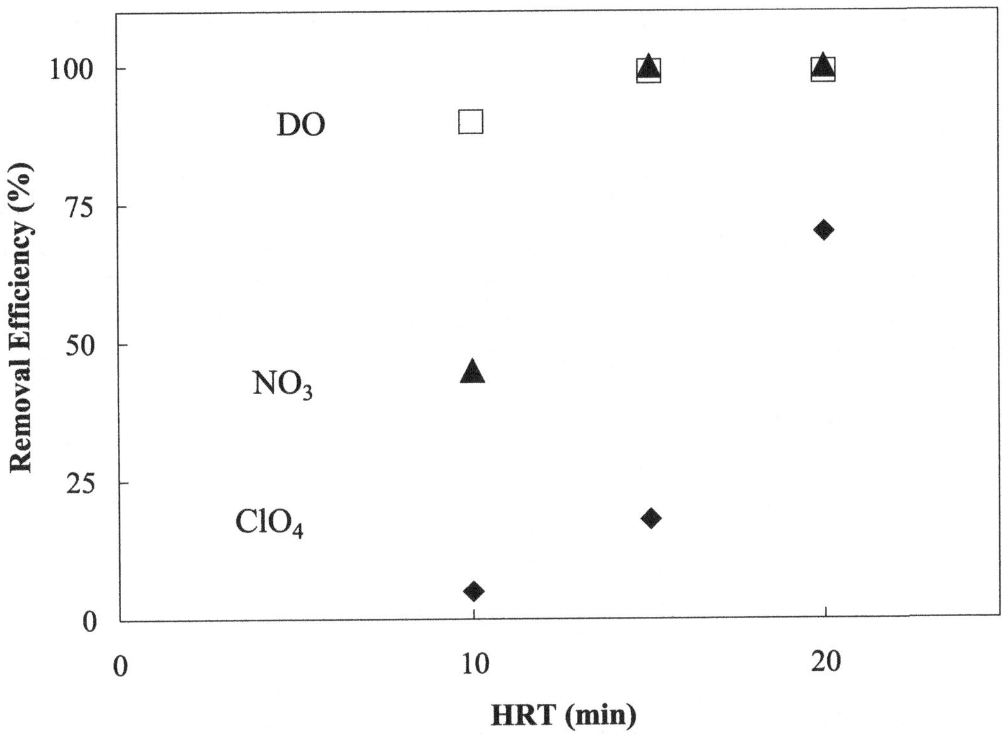

Figure 7.14 Influence of HRT on M6 electron acceptor removal efficiency

operating conditions were applied to the system, and the system performance was evaluated with respect to:

- Hydraulic Retention Time
- Effectiveness of Air Scour Cleaning

Hydraulic Retention Time

After operating the M6 reactor to steady-state conditions with an HRT of 20 minutes, the flowrate was increased to evaluate system performance as different HRTs. Figure 7.14 shows the reactor performance at different HRTs. At an HRT of 20 minutes, the system exhibited performance similar to that of M20 with respect to perchlorate removal (70 percent), while oxygen and nitrate were reduced to below analytical detection limits. Similar to results observed during operation of the M20 reactor, nitrate breakthrough was observed at an HRT of 10 minutes. Consequentially, the perchlorate reduction rate also dramatically suffered due to the fact that the excess nitrate appeared to inhibit perchlorate reduction.

Effectiveness of Air Scour

One of the features of this modified module was the addition of improved reactor cleaning using four independent air scour ports, instead of an air scour port common with the influent water. Similar to the modification of influent and effluent water ports, air scour ports were created to provide more efficient and uniform cleaning of the fibers. In addition, a separate large drain port was available for removal of solids. The availability of several ports for independent air scour, feed water, and drainage allowed the project team to demonstrate and improved an effective means of cleaning the MBfR.

Air scour cleaning was performed daily with no decrease in reactor performance (i.e. system recovery within 3 HRTs). In addition, biofilm was visually observed to be thinner and denser that what was observed before in the previous MBfR designs. The thin and dense biofilm achieved during operation the M6 reactor was critical for maintaining good system hydraulics and mass transfer of nutrients to the biofilm, thereby eliminating the propensity for fibers to form biofilm overgrowth, promoting the clumping of fibers.

HYDRAULIC EVALUATION—TRACER TESTING

To further characterize the hydraulics of the MBfRs under selected operational conditions, a series of tracer tests was performed to identify and confirm key factors required for the successful design and operation of an MBfR.

Tracer Testing Methodology

Sodium chloride was used as a conservative tracer in all tests. Following a step-input of tracer, samples of the effluent were collected for a minimum of four theoretical hydraulic retention times (T) to ensure that sampling was long enough to allow the effluent to reach the same concentration as the influent. Once samples were collected, chloride ion (Cl^-) concentrations were analyzed by ion chromatography and plotted to produce the normalized response (C/Co) versus time (F-Curve). Further analysis of the step-input tracer tests data using the residence time distribution (RTD) function (E-Curve) revealed further details about system hydraulics including: the mean residence time (t_m) and the extent of axial dispersion or dispersivity ($D/\mu L$).

Tracer Results

Tracer tests were performed with two reactor designs: the M3 and M20 reactors. Experiments were specifically designed to evaluate reactor hydraulics under the following conditions:

- Impact of reactor cleaning
- Presence or absence of recycle flow
- Low versus high influent flow

Table 7.1 summarizes results from all tests performed during the hydraulic evaluation.

Impact of Reactor Cleaning

The effectiveness of reactor cleaning was characterized by performing a tracer test using the M3 design module before and after a regular mild acid cleaning. For both tests, the system was operated at a flowrate of 0.15 gpm (theoretical HRT = 30 min) and without recycle flow. Without recycle flow, the system roughly approximated a plug flow reactor. The mean residence times

Table 7.1
Hydraulic parameters for tested designs

Condition	R (gpm)	T (min)	t_m (min)	t_m/T	D/μL
M3 Reactor					
Pre acid clean	0	30	36	1.20	0.16
Post acid clean	0	30	35	1.16	0.10
Low flow w/ recycle	30	30	37	1.23	0.43
High flow w/ recycle	13	15	7	0.47	0.26
M20 Reactor					
Low flow w/o recycle	0	30	16	0.53	0.02
Low flow w/ recycle	25	30	28	0.93	2.46
High flow w/ recycle	13	15	17	1.12	0.36

calculated from the RTDs for both conditions (pre = 36 min and post = 35 min) exceeded the theoretical HRT of 30 minutes. This is a result of hydraulic "dead-zones" within the reactor, which can be seen through the significant tailing of the RTD (Figure 7.15).

As shown in Figure 7.15, a significant difference was noticeable in the shape of the corresponding RTD curves. Prior to the acid cleaning, the effluent response exhibited a significant amount of dispersion, as seen in the broadness of the curve. However, after a mild acid cleaning was performed, the effluent response was improved, better resembling a PFR. This was confirmed by calculating the dispersivity of the tracer response. Before cleaning, D/μL was 0.16 while D/μL decreased to 0.10 after cleaning. For ideal plug flow conditions D/μL = 0. Additionally, small and intermediate amounts of dispersion are observed when the dispersivity is 0.002 and 0.025, respectively (Levenspeil, 1972). Thus the M3 reactor had significant dispersion before and after mild acid cleaning.

Presence or Absence of Recycle Flow

Previously, the use of recycle flow was an important factor for maintaining mass transport and improving hydraulics. Tracer tests were performed with both types of reactors (M3 and M20) to characterize the improvement.

For the M3 reactor, tracer tests were performed when the system was operated at flowrate of 0.15 gpm (30 min HRT) with recycle ratios of R=0 and R=30. As shown in Figure 7.16, when recycle flow was used, the effluent response roughly approximated a CSTR. The D/μL was calculated to be 0.43 when recycle flow was used. Compared to the same system without recycle, the dispersivity demonstrated the impact of mixing as a result of the recycle flow. However, even with a high degree of mixing, the mean residence time in excess of the theoretical HRT revealed that hydraulic dead zones persisted.

Similarly, a tracer test was performed on the M20 reactor operated at a flowrate of 0.12 gpm (theoretical HRT = 30 min) with recycle ratios of R=0 and R=25. The RTD response of the M20 was

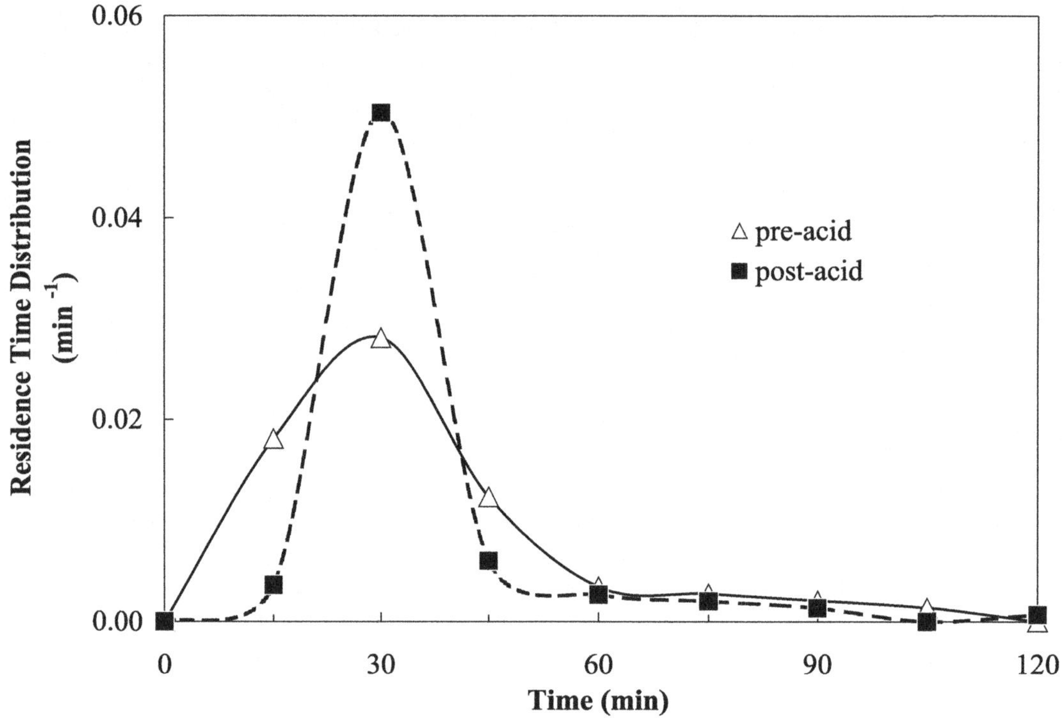

Figure 7.15 Residence time distribution of the M3 reactor before and after acid cleaning (Q=0.15 gpm, R=0)

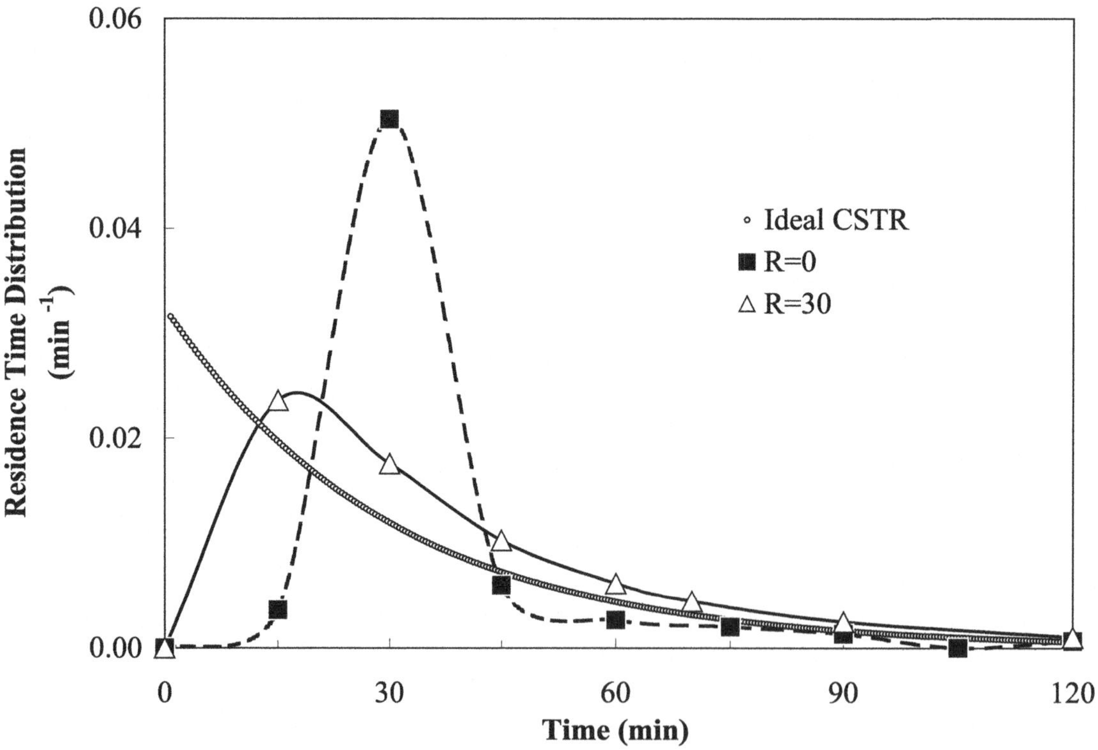

Figure 7.16 Impact of recycle flow on the M3 reactor (Q=0.15 gpm, R=0 and R=30)

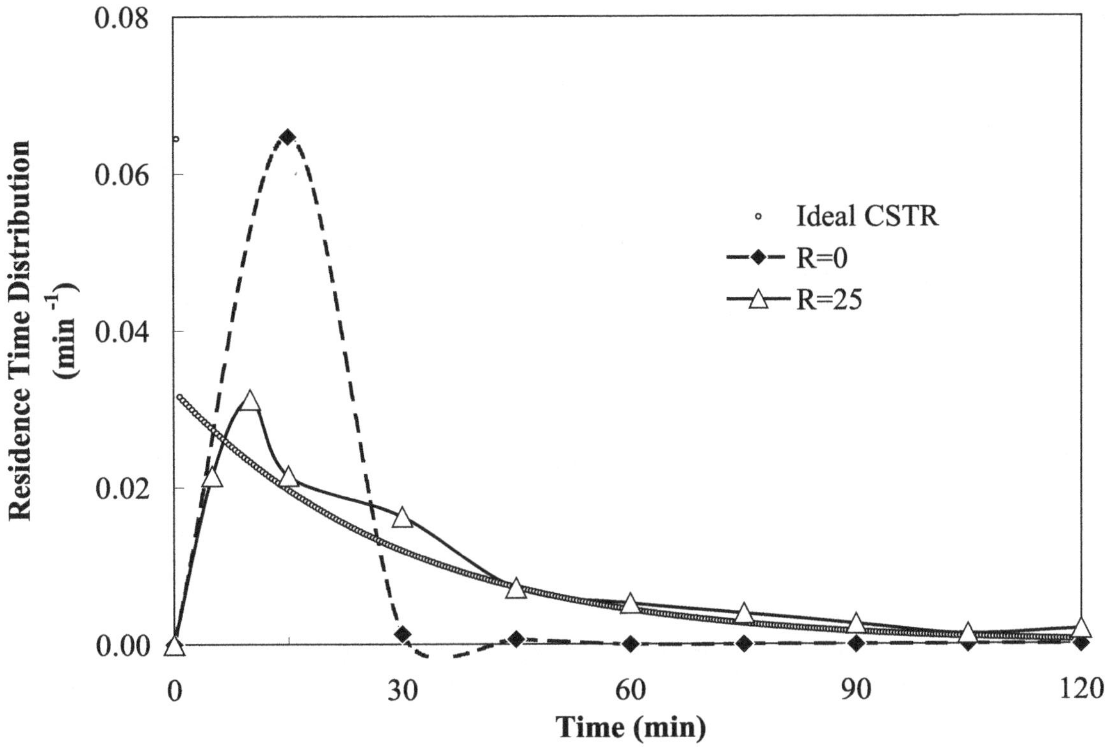

Figure 7.17 Impact of recycle flow on the M20 module (Q=0.12 gpm, R=0 and R=25)

slightly different than what was previously observed with the M3 design. With R=25, the calculated t_m of 28 minutes correlated well with the theoretical HRT. In addition, the RTD response approaches that of a CSTR with the high recycle, as seen by the E-curve in Figure 7.17. The dispersivity of this condition was calculated as $D/\mu L=2.46$, indicating nearly ideal mixed flow conditions. Without recycle, however, t_m was calculated to be only 16 minutes with little dispersion, $D/\mu L = 0.02$. $t_m/T = 0.53$ indicates serious short-circuiting.

Low Versus High Influent Flow

One of the major limitations of the MBfR designs tested was the inability to maintain perchlorate reduction under high flow conditions. This was hypothesized to be a result of a combination of both channeling occurring within the reactor, resulting in short-circuiting effects, and limited biodegradation time. To evaluate the effect of low versus high flow conditions in the M3 reactor, tracer tests were performed with flowrates corresponding to theoretical HRTs of 30 and 15 minutes. Based on the results of previous tracer tests, recycle flow was also used. As seen in Table 7.1, at a low flow condition (theoretical HRT = 30 min), the calculated t_m was 37 minutes yielding a t_m/T of 1.23. However, at the high flow condition (theoretical HRT = 15 min), t_m was 7 minutes, resulting in $t_m/T = 0.47$. Under this condition, the recycle flow was insufficient to ensure adequate mixing throughout the reactor to avoid serious short-circuiting.

Similar tests were run for the M20 reactor at flowrates corresponding to theoretical HRTs of 30 and 15 minutes. Recycle was also used in both tests. In contrast to what was observed in the M3 reactor, the calculated t_m values were close to the theoretical HRTs. The t_m/T ratio was 0.93 and 1.12 for the low and high flow conditions, respectively.

Hydraulic Testing Conclusions

An ideal MBfR reactor would have a high membrane surface area/packing density (for delivery of hydrogen and subsequent reduction of perchlorate) and would be capable of operating with high flowrates without serious short-circuiting. The success of these two features would only occur if the reactor was hydraulically sound. Two extremes were evaluated during the tracer tests: the M3 (3 percent packing density) and M20 (20 percent packing density) reactors.

As a part of the O&M requirements for operation of the MBfR, acid cleaning is recommended on a monthly basis to effectively clean the reactor housing and membrane fibers of extraneous biomass overgrowth that may accumulate during operation. Tracer tests performed prior to and after acid cleaning, revealed that mild acid cleaning was an effective means for restoring the hydraulic performance of the MBfR. Given the design of the M3 and M20 reactors, it was also evident that the recycle flow improved mass transfer throughout the reactors and reduced short-circuiting.

The tracer tests were performed to demonstrate the impacts of the MBfR designs on hydraulics. They may not, however, represent long-term operational hydraulic performance. Although the M20 reactor demonstrated good hydraulics under low and high flow conditions while the reactor was clean, it is important to note that cleaning efficiency was poor due to the high fiber packing density. Additional modifications and hydraulic evaluations through computational fluid dynamics or tracer studies could be used to further optimize the design based on: packing density/membrane surface area, reactor configuration (i.e. baffling, flow entry, flow exit, etc.), and recycle requirement. However, actual operation would be required to asses the impacts of the biofilm on long term operability and performance.

CHAPTER 8
POTABLE WATER PRODUCTION REQUIREMENTS

The MBfR was found to effectively remove perchlorate to below the 4 μg/L CaDHS action limit. However, additional development is required to further optimize the process. The overall goal of the MBfR is to treat groundwater impaired with perchlorate for drinking water use. To achieve this goal, the project team identified three major requirements that would be necessary for implementation of MBfR technology used for the production of potable water:

- Post-MBfR unit operations required for further treatment before distribution
- Compliance with current water quality regulations
- Comments from state and local agencies regarding regulatory approval requirements

SUBSEQUENT UNIT OPERATIONS

While perchlorate and other electron acceptors (i.e., oxygen and nitrate) were reduced in the MBfR, additional processes were necessary to complete the pilot-scale treatment system. Following the MBfR, an aeration process was used to achieve two primary goals: first, it oxygenated the water in preparation for its introduction into a distribution system as a drinking water source; and second, it provided sufficient oxygen for operating the downstream media filter in an aerobic biodegradation mode to remove biodegradable organic matter.

While aerobic activity was important in the media filter, the filter's primary role was to remove any biomass that may detach during operation of the MBfR; excess biomass within the MBfR itself was regularly removed by air scour. This biomass-laden water, however, was directed to waste and was not allowed to pass onto the media filters. In a full-scale operation, the media filter effluent would then be dosed with chlorine to provide disinfection and carry a residual through the distribution system. Alternatively, media filtration could be replaced by membrane filtration for increased protection against the breakthrough of sloughed biomass.

Aeration Process Operation

Treated effluent from the MBfR was diverted to the pilot-scale aeration basin. Fine-bubble diffusers located in the bottom of the basin continuously sparged air to oxygenate the water. After aeration, the dissolved oxygen concentration of the treated effluent was typically 6 to 7 mg/L. Additionally, the dissolved hydrogen levels entering the aeration basin were typically 300-500 μg/L. Aeration effectively stripped out the residual hydrogen such that the effluent concentrations were below detection limits (5 μg/L) after aeration.

Media Filtration Operation

The dual media filter was operated at a filter loading rate of 2 gpm/sq-ft. Turbidity was continuously monitored and averaged 0.5 NTU in the filter influent. Filtered effluent was consistently measured at 0.05 NTU. The headloss through the filter column was monitored as the sloughed off biomass was removed by the bed of filter media. As shown in Figure 8.1, the buildup of headloss was gradual over several days. As the filter headloss approach 100 inches of water, the

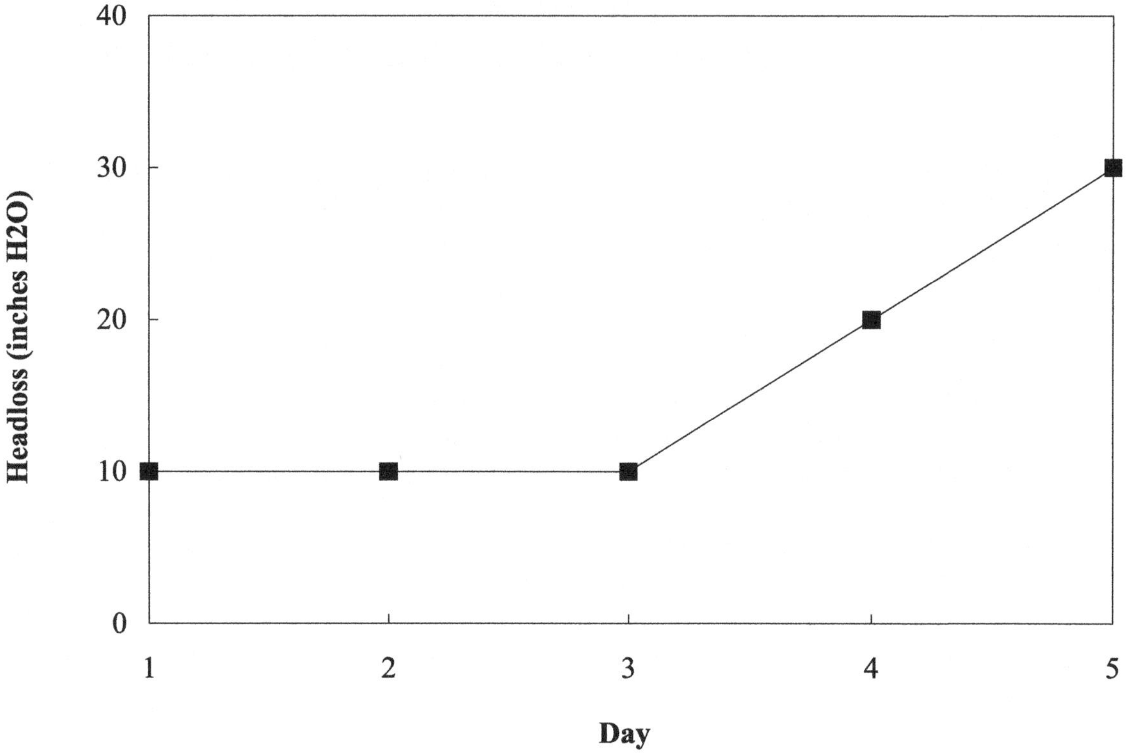

Figure 8.1 Media filter headloss

system was backwashed with air scour. The backwash frequency observed while the filter was in continuous operation was typically 7 to 10 days.

MBfR TREATMENT SYSTEM WATER QUALITY

An extensive sampling campaign was performed to characterize the water quality through of the entire pilot-scale treatment process. All treated effluent water quality parameters were found to be in compliance with both the National Primary Drinking Water Standards and Secondary Drinking Water Regulations. Additional analyses were performed to determine if other compounds were either reduced or produced through the treatment train. These additional analyses included: SVOCs, VOCs, aldehydes, aldicarbs, diquat/paraquat, TOC/BDOC, HPCs, and the TTHMFP. As listed in Table 8.1, the only significant changes were detected in the HPCs. As expected, significant concentrations of bacteria were measured in the effluent of this biological treatment process. However, dosing preformed chloramines (2.5 mg/L) was sufficient to reduce the HPCs to zero.

REGULATORY REQUIREMENTS

The project team understands the importance of input from local and state agencies responsible for potential regulatory approval of the MBfR treatment process. Although the MBfR technology was in its development stages, the teams of regulators from the following state health

Table 8.1
Summary of MBfR process treated water quality

Parameter	Method	Comment
Primary Drinking Water Standards	Various	In compliance
Secondary Drinking Water Standards	Various	In compliance
Additional testing		
Semivolatiles	EPA 525.2	ND
Regulated VOCs	EPA 524.2	ND
Aldicarbs	EPA 531.1	ND
Diquat / paraquat	EPA 549.2	ND
HPCs	SM 9215B	Raw water (1600 CFU/mL) MBfR effluent (>5700 CFU/mL) Media filter effluent (>5700 CFU/mL) Post-Cl_2 addition (2.5 mg/L NH_2Cl; 0 CFU/mL)
TOC / BDOC	SM 9215B Joret, Levy, and Volk (1991)	Raw water (<0.5 mg/L) MBfR effluent (0.6 mg/L) Media filter effluent (<0.5 mg/L) Post-Cl_2 addition (<0.5 mg/L)
TTHMFP	SM5710	Post-Cl_2 addition (<10μg/L)

departments were contacted to obtain their comments and/or concerns with the MBfR process for the production of potable quality water:

- California
- Utah
- Texas
- Massachusetts

The following summarizes comments of regulators from California, Texas, and Massachusetts. Their formal responses are included in Appendix B for reference.

- Site-specific issues
  - Avoid variability in flow
  - Avoid variability in water quality (i.e., seasonal and temporal changes)
  - Characterize source specific hydrogen feed rates
  - Organic contaminants in the source water may require independent downstream unit treatment process

- MBfR reactor specific comments
  - Identification of microorganisms is not required since indigenous organisms are used
  - All chemicals should be ANSI/NSF 60 certified
  - Reactor components in contact with source water must be ANSI/NSF 61 certified
  - Develop a reliable feed-forward hydrogen control
  - On-line ClO_4^-/NO_3^- monitoring systems should be incorporated
  - Periodic integrity testing is suggested for detecting broken fibers

- MBfR process comments
  - May need to meet disinfection requirements identified in the Surface Water Treatment Rule
  - Contact other regulatory bodies to address worker health and safety issues with hydrogen use and storage
  - Characterize downtime required for service and maintenance (i.e., air scour, acid cleaning, etc.)
  - To meet the multiple barrier requirement, media and membrane filtration needs to be investigated as well as a second barrier capable of at least 4-log removal of bacteria and virus
  - MBfR wastewater effluent quality must be characterized to evaluate any request for a discharge permit.

Overall, comments from the regulators were positive and informative; many of which have already been addressed in the current scope of research presented in this report. Dr. Rick Sakaji (CaDHS) found that the, "membrane bio(film)reactor does appear to be a promising technology for perchlorate reduction. The Department looks forward to future communications from you regarding this technology. We hope these comments provide you with the feedback you are seeking and help direct future research so that the regulatory and permitting issues raised, can be addressed." Additional comments, not considered in this research will addressed in the future development of MBfR technology and will be reviewed and discussed with all appropriate regulatory agencies.

CHAPTER 9
ENGINEERING ANALYSIS

Extensive bench- and pilot-scale testing clearly demonstrated that a hydrogen-fed membrane biofilm reactor can be used to treat perchlorate-contaminated waters. The objective of this chapter was to determine if this process is economically feasible for full-scale application. The economic feasibility evaluation provides a set of projected capital and operation and maintenance (O&M) requirements for a full-scale MBfR treatment plant. The costs presented here are for a complete treatment system including: the MBfR, post-aeration treatment, multimedia filtration, construction and start-up related costs, and annual operation and maintenance cost.

A cost model was developed to determine the impact of the various design factors effecting capital, operation, and maintenance costs. This model was derived using input variables obtained from field results and theoretical considerations for each design. The objective of running the model was to:

1. Determine if the full-scale MBfR plant is cost competitive with existing perchlorate removal facilities,
2. Quantify the impact of the MBfR design on the full-scale treatment costs, and
3. Evaluate MBfR costs under improved (ideal) design and operating conditions (referred to as projected MBfR design for the remained of this report).

COMPARISON TO EXISTING FULL-SCALE TREATMENT

In order to determine if the full-scale MBfR plant is cost competitive with other perchlorate removal technologies it was compared to an existing perchlorate removal facility at the La Puente Valley County Water District (LPVCWD). The scenarios evaluated were:

- LPVCWD ion exchange facility with brine line waste disposal (existing),
- LPVCWD ion exchange facility with brine treatment (estimated),
- MBfR M20 design, and
- Projected MBfR design.

Table 9.1 summarizes results obtained from the cost model for each of these scenarios. The following sections outline the designs evaluated and discuss the capital, operation and maintenance costs in detail used for this cost comparison.

As shown in Figure 9.1, there is a wide range in the unit water production treatment costs associated with the scenarios evaluated. Perchlorate treatment at the existing LPVCWD facility costs $1.09/1000 gal ($356/acre-ft). Currently, this system removes the perchlorate and discharges it into a dedicated brine line. However, if perchlorate-laden brine discharge were no longer allowed, the plant would need to provide a perchlorate destruction or removal mechanism prior to disposal. The addition of a Perchlorate Nitrate Destruction Model (PNDM, Calgon Carbon Corporation) increases the current cost per unit production by 21 percent, approximately $1.32/1000 gal ($430/acre-ft), primarily attributed to a significant increase in capital costs. Both ions exchange methods are considerably lower than the cost calculated for the initial MBfR

Table 9.1
Full-scale IX plants versus full-scale MBfR plant

Process	Annual O&M cost ($/acre-ft)	Amortized capital costs ($/acre-ft)	Total treatment costs ($/acre-ft)	($/1000 gal)
LPVCWD (IX only)	$261	$95	$356	$1.09
LPVCWD (IX with brine treatment)	$257	$173	$430	$1.32
MBfR—M20 design	$1,900	$2,400	$4,300	$13.20
MBfR—projected design	$140	$220	$360	$1.10

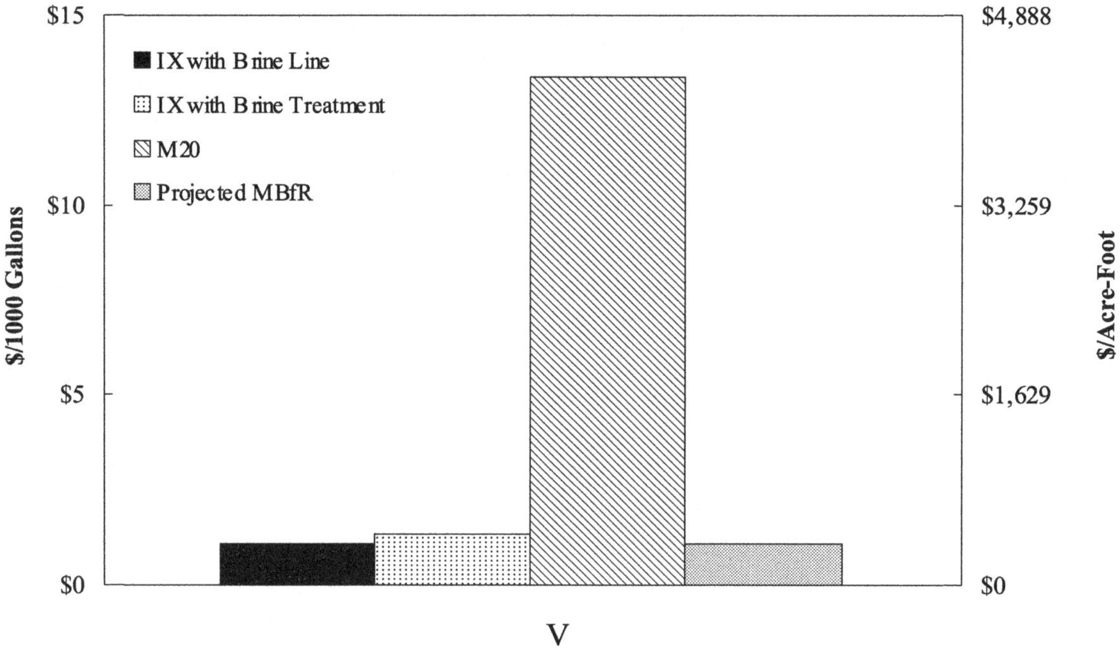

Figure 9.1 MBfR scale-up (2500 gpm) versus existing full-scale ion exchange

design. However, based on the projected modifications, it was anticipated that the project MBfR costs are approximately $1.10/1000 gal ($360/acre-ft), presenting a potentially cost-competitive treatment alternative.

LA PUENTE VALLEY COUNTY WATER DISTRICT ION EXCHANGE FACILITY

The La Puente Valley County Water District (LPVCWD) currently uses a 2,500-gpm Calgon Carbon Corporation (Calgon) ISEP ion exchange system for perchlorate and nitrate removal. The ion exchange system is preceded by an air stripper for VOC removal and followed by an ultraviolet (UV) treatment system for NDMA removal. To maintain consistency when comparing full-scale ion exchange to MBfR technology, the capital and O&M costs associated with the air stripper and the UV system were removed from the total system costs.

Table 9.2
LPVCWD (2500 gpm) estimated annual O&M costs

O&M parameter	No brine treatment	PNDM brine
Salt	$406,000	$75,000
Maintenance	$166,000	$250,000
Other consumables	$150,000	$227,000
Power	$252,000	$335,000
Utilities (water + gas)	$2,500	$58,000
Labor (+overhead)	$45,000	$90,000
Brine disposal	$33,000	$2,500
Total cost per year	$1,054,500	$1,037,500
Cost per acre-foot	$261/AF	$257/AF
Cost per 1000 gallons	$0.80	$0.79

The ion exchange process produces a concentrated perchlorate- and nitrate-laden waste stream during the regeneration process. The perchlorate in the waste brine is not destroyed, but simply disposed of through a dedicated brine discharge line. Future regulations may prevent the use of a brine line for discharging perchlorate-laden waste brine. With brine treatment being a possible future reality, the LPVCWD ion exchange system was evaluated based on both the current operating conditions and with estimated brine treatment costs. Costs associated with brine treatment were based on estimates provided by Calgon for their Perchlorate and Nitrate Destruction Module (PNDM). The PNDM is the brine treatment component of the Calgon ISEP+ system which treats and reuses spent brine from the ion-exchange regeneration process.

Annual O&M costs for the ion exchange system (no brine treatment) were provided by LPVCWD and Calgon. This data is presented in Table 9.2 and represents the costs associated with a production rate of 2500 gpm. The O&M costs do not include LPVCWD associated costs for water quality testing or other engineering related costs due to systems unrelated to the ion exchange system. The estimated capital costs for the 2500 gpm LPVCWD perchlorate removal system are presented in Table 9.3. Capital costs are presented for the ion exchange system, with and without brine treatment.

ANTICIPATED FULL-SCALE MBfR TREATMENT COST TRENDS

The pilot-scale MBfRs were based on adaptations of membrane modules used for particle separations from water. The results from the pilot studies showed that the pilot designs were far from optimized in terms of hydraulics, mass transport, and cleaning. Although much-improved designs clearly are possible, the results from the pilot studies provide a basis for identifying key trends in what control system costs and, therefore, what aspect of design deserve the most intense attention. Therefore, full-scale treatment cost projections (based directly on the pilot results) were calculated from a combination of standard engineering practice and the pilot results. The assumptions used in these projections are presented in this section.

Table 9.3
LPVCWD (2500 gpm) estimated capital and construction related costs

Capital cost parameter	No brine treatment	Brine treatment
Brine line connection	$40,000	$0
ISEP	$2,200,000	$2,200,000
ISEP installation	$770,000	$770,000
PNDM	$0	$2,000,000
PNDM installation	$0	$700,000
Facility structure	$240,000	$290,000
Construction related costs	$1,200,000	$2,100,00
EL&A costs*	$355,000	$650,000
Total cost to construct	$4,805,000	$8,710,000
Amortized capital cost	$383,000/yr	$702,000/yr
Amortized cost per acre-foot†	$95/AF	$174/AF
Cost per 1000 gallons	$0.30	$0.53

* Engineering, legal and administrative costs
† 2500 gpm system at 5% interest for 20-years.

The MBfR system was evaluated for each of the designs: M3, M6, and M20. Each design modification had an impact on the system performance. These changes in performance were evaluated in the cost model to determine if there was an effect to the overall cost. The major components considered in each evaluation included:

- Design flow rate
- Water quality
- MBfR design data
- Capital costs
- Operation and maintenance (O&M) costs
- Construction related costs

Detailed information on the input variables used for the cost model can be found in Appendix A.

MBfR Capital Cost

Capital costs are those items that contribute to a new MBfR facility, including treatment equipment, installation, and other construction-related costs. The major components of the capital costs associated with a new MBfR plant and the estimated percentage of the total cost to construct are:

Treatment Related Equipment	46%
Construction Related Costs	24%

Engineering, Legal and Administrative	17%
Project Risk Factor	13%

The equipment contributes the most to the capital costs. Of the equipment costs, 95 percent are directly related to the number of reactors required. This includes the initial reactors, related support equipment, and installation of the equipment directly related to the MBfR process. The capital cost does not include land acquisitions.

In addition to perchlorate and nitrate being removed, the MBfR process removes oxygen from the water. Therefore, aeration equipment is required downstream from the bioreactor. The addition of a conventional aeration system will introduce dissolved oxygen into the water supply, but will also cause the system to break head; this will require an effluent pump(s) to boost the pressure to required levels prior to the distribution system. Additional capital requirements include multimedia filters to remove detached biomass downstream of the MBfR, a pre-strainer to prevent suspended solids from entering the MBfR, and a structure to house the MBfR plant. Individual construction related costs and assumptions can be found in Appendix A.

MBfR Annual O&M Costs

Annual O&M costs are associated with daily consumable variables. These consumables include utilities as well as daily maintenance requirements. For this cost analysis, the membrane life was assumed to be 5 years, with all other equipment life at 20 years. Additional details related to the O&M related costs can be found in the Appendix A. The major components of the O&M costs included:

Membrane replacement	70%
Maintenance	18%
Membrane Cleaning	9%

MBfR Reactor Designs

The three reactor designs tested at pilot-scale (M3, M6, and M20) were used in the cost model to:

1. Identify trends that make and MBfR design more or less economical
2. Identify the reactor parameters have the most influence on MBfR costs
3. Determine the impact of economy of scale

Effects of Reactor Design

There is a notable difference in the estimated full-scale cost between the three MBfR designs. Figure 9.2 presents the cost comparison between the three designs for a 2500 gpm system. The M20 design has the least cost per unit production of the three designs, 37 percent less than the M3 design. The cost reduction is directly related to reduced number of reactors required for the M20 MBfR design compared to the M3 and M6 designs. The higher specific surface area of the M20 design was the dominant factor providing the cost advantage.

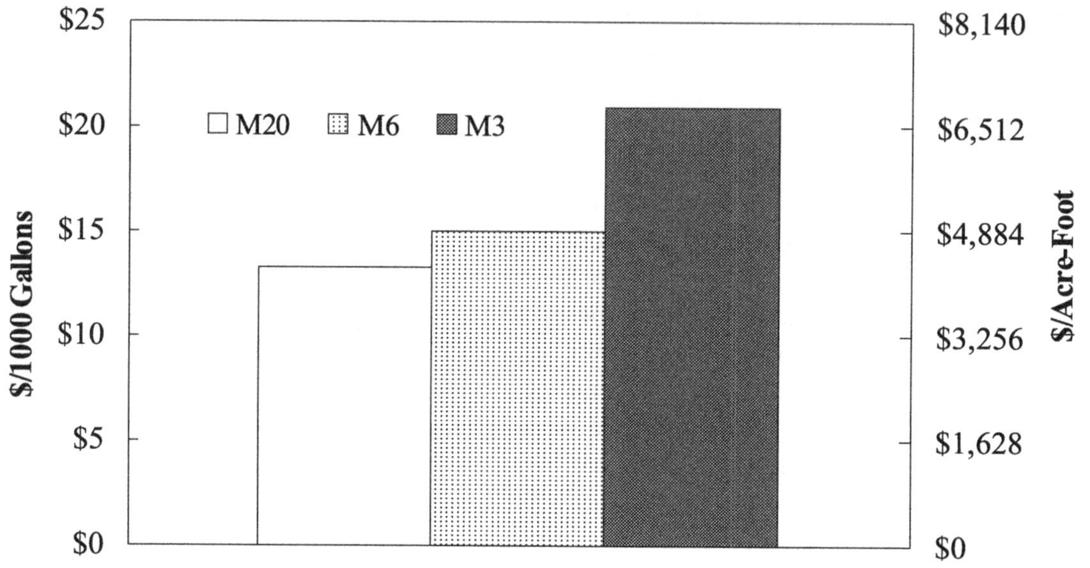

Figure 9.2 Total system cost per unit production per reactor design (2,500 gpm system)

Table 9.4
Total system cost per reactor design ($/AF)

	M3	M6	M20
Amortized capital*	$3600	$2600	$2400
Annual O&M	$3200	$2200	$1900
Total cost ($/1000 gal)	$6800 ($21.00)	$4800 ($15.00)	$4300 ($13.20)

Cost estimates rounded to the nearest $100.
*Amortized capital based on 2500 gpm system operating 365 days per year, financed for 20 years at 5 percent interest.

The costs information presented in Figure 9.2 are based on the amortized capital costs plus the annual operations and maintenance costs. The data that corresponds with this figure is presented in Table 9.4 and is discussed in more detail in the following sub-sections.

Parameters Affecting MBfR Costs

Based upon the reactor comparison presented above, it was established that the packing density of the reactor had a significant impact on the overall system cost. The pilot studies were conducted using 3, 6, and 20 percent packing densities. To evaluate these and other packing densities, two assumptions were made:

1. A relationship exists between packing density (surface area) and flux, and
2. This relationship can be estimated based on the three MBfR pilot designs, their fluxes and membrane surface areas.

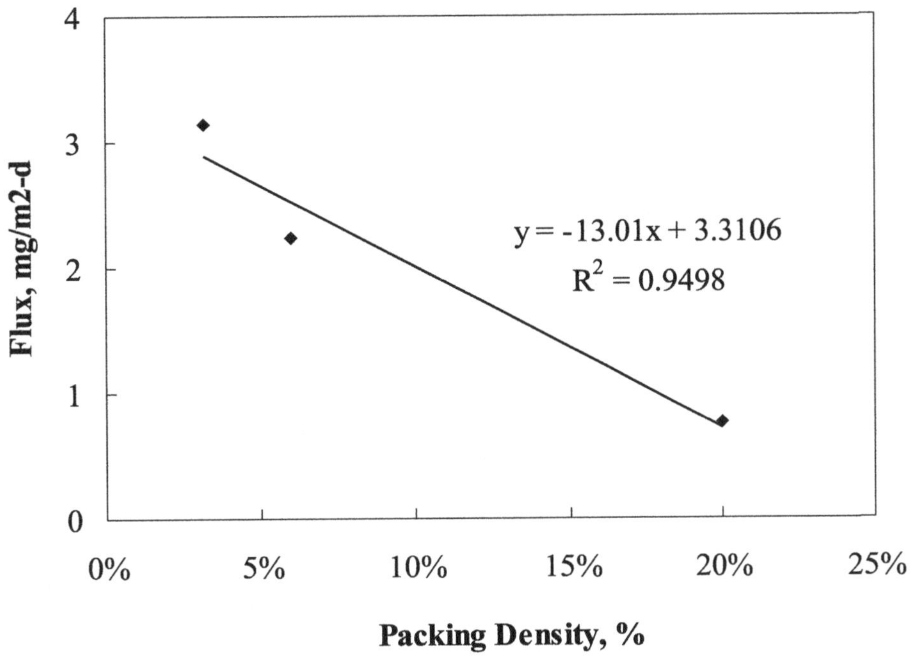

Figure 9.3 Flux versus packing density

Given these assumptions, the flux and packing density were plotted for each MBfR design, as shown in Figure 9.3. From this graphic, a formula was established for a potential relationship between packing density and flux. Using this relationship, flux was calculated for a range of packing densities. These were then used to calculate the required number of reactors and, ultimately, the cost per unit production of water.

Using this approach, as shown in Figure 9.4, the optimal packing density is in the range of 9 percent to 15 percent.

PROJECTED MBfR

The pilot MBfR systems were far from optimal, since they were based on adaptations of membrane-filtration units. Nevertheless, the pilot-scale testing demonstrated that significant improvements could be made with simple modifications. Focused attention to improving the process configuration of this emerging process should result in major reductions in overall treatment costs. Utilizing the results from the bench- and pilot-scale studies, as well as good engineering concepts, we evaluated a projected MBfR design that overcomes the limitations of the existing pilot systems. The projected design has these features:

- The perchlorate flux is 23 mg/m^2-day, which is the highest flux obtained in the bench-scale studies with an influent of 100 μg/L of perchlorate. An increased flux decreased the required membrane surface area proportionally.
- Each module is increased is size by containing multiple submerged elements. Having multiple elements reduced fabrication costs.
- Making the fibers longer, also to reduce fabrication costs.

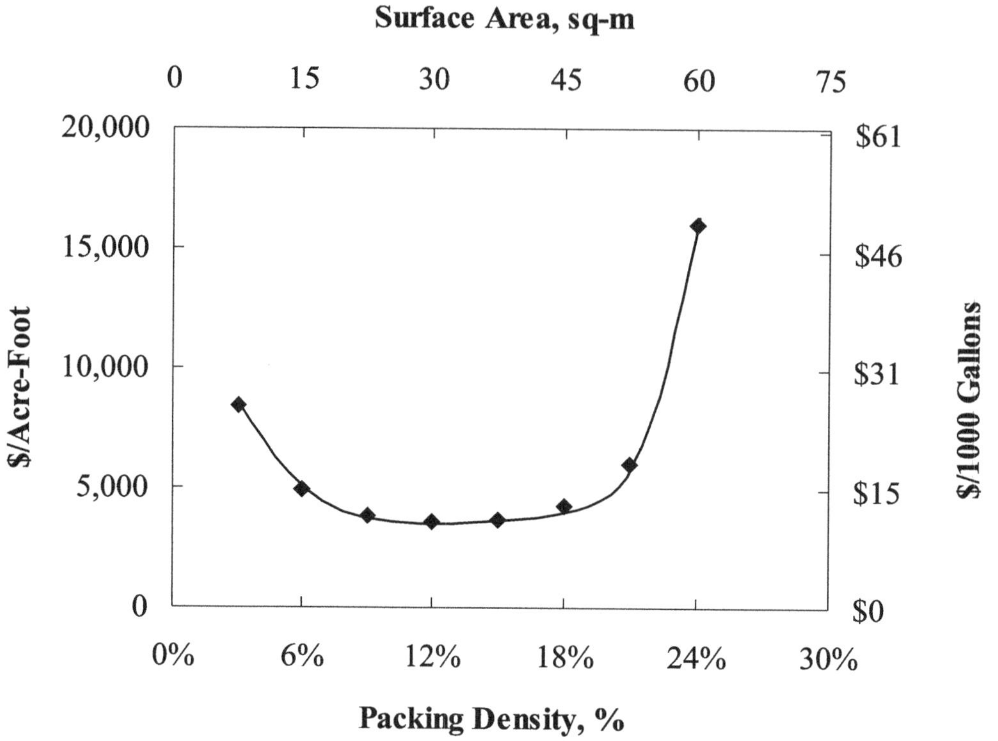

Figure 9.4 Effects of packing density on reactor costs

Table 9.5
Design comparison of M20 to the projected MBfR

Design parameter	M20 MBfR	Projected MBfR
Reactor configuration	Individual modules	Multiple elements within a reactor
Flux for 100 μg/L ClO_4^- influent, mg/m^2-day	0.8	23
Membrane surface area required, m^2	1,000,000	57,000
Individual module/element surface area, m^2	52.8	66.1
Total number of module/element	19,000	860
Reactor cross sectional area, ft^2	0.17	1.75
Number of fibers per reactor	50,000	44,000
Reactor height, ft	4	6
Fiber length, ft	3.6	5.6
Fiber OD, μm	280	280

Table 9.6
Capital costs comparison of the projected MBfR design to M20 (2500 gpm)

Capital cost parameter	M20 MBfR	Projected MBfR
Initial membrane cost	$23,000,000	$700,000
Membrane related support equipment cost	$12,000,000	$525,000
Auxiliary equipment cost	$1,500,000	$1,500,000
Facility cost	$2,600,000	$712,000
Construction related cost	$25,000,000	$2,100,000
EL&A cost*	$21,000,000	$1,700,000
Project risk factor cost	$16,800,000	$1,400,000
Total capital costs	$122,000,000	$10,200,000
Amortized capital cost†	$9,800,000/yr	$820,000/yr
Capital cost per acre-foot	$2400/AF	$200/AF
per 1000 gal	$7.35	$0.70

* Engineering, Legal and Administrative
† Amortized for a 2500 gpm system, operating 365 days per year, at a 5% interest and 20-year loan

Table 9.5 compares the design conditions of the M20 pilot MBfR and the projected MBfR design. By achieving the improvements noted above, the total number of reactors required significantly decreased from 19,000 to 860.

As previously identified, the reactor costs are a significant percentage of the capital and the annual O&M costs. By reducing the number of reactors, the overall cost subsequently decreases. The associated capital costs for the projected design are presented in Table 9.6.

Reducing the number of reactors has a domino effect on the remaining cost factors, as seen in Table 9.6. Having fewer reactors, means less support equipment, a smaller facility footprint, and less construction, EL&A, and overall risk factor costs translating into a lower total production cost for the projected MBfR. The unit production cost for a 2500 gpm system reduces from $2400/AF to $200/AF in the projected MBfR design.

The associated annual O&M costs also decreases with a decrease in the number of reactors required for the system. As presented in Table 9.7, the O&M cost of the proposed MBfR design is lower than the M20 design primarily due to lower membrane replacement, cleaning, and maintenance.

By decreasing the number of reactors, the projected MBfR design has a reduced capital and O&M costs associated with the total system. This reduces the life cycle cost from $4300/AF ($13.20/1000gal) to $360/AF ($1.10/1000gal), thus making the MBfR both technically feasible and economically competitive.

DISCUSSION

The bench- and pilot-scale data revealed that the MBfR process is technically feasible for the removal of perchlorate. However, the objectives of the engineering analysis section were to:

Table 9.7
O&M costs comparison of the projected MBfR design to M20 (2500 gpm)

Annual O&M cost parameter	M20 MBfR	Projected MBfR
Annual hydrogen cost	$23,000	$23,000
Annual membrane replacement costs*	$5,700,000	$140,000
Membrane cleaning cost	$622,000	$135,000
Energy cost	$167,000	$129,000
Labor cost	$37,000	$37,000
Maintenance cost	$1,200,000	$92,000
Total annual O&M costs	$7,650,000	$556,000
Annual O&M cost per acre-foot production†	$1900/AF	$140/AF
per 1000 gal	$5.85	$0.40

* Assuming a 5-year membrane life, 20% replaced per year
† Estimated for 2500 gpm system

1. Determine if the full-scale MBfR plant is cost competitive with existing perchlorate removal facilities,
2. Determine the impact of various MBfR design factors on the full-scale costs, and
3. Evaluate MBfR costs under improved design operating conditions.

Upon evaluation of the existing MBfR designs, it was discovered that design has a large impact on the life cycle cost of the system. This was evident already between the M3 (first generation) and the M20 designs where a 37 percent reduction in life cycle cost was observed for a 2500 gpm system. Further investigation through the use of the cost model revealed that additional design modifications would significantly reduce the life cycle cost of the MBfR. The conceptual-level cost projections for the projected MBfR system are anticipated to be cost competitive with other existing perchlorate removal process. However, additional investigation including pilot-scale testing of the proposed design modifications are recommended.

There appears to be little economy of scale for the current MBfR design, due to the large initial cost of the system. Over the range of 500 to 2500 gpm, a 10 precent reduction in life cycle cost was realized for the M20 design. While minimal, a reduction in cost is noticed for economy of scale.

CHAPTER 10
SUMMARY

Perchlorate can be a significant challenge to water utilities, as it is not removed by conventional water treatment, and advanced water treatment processes are expensive. While perchlorate can be biologically reduced to innocuous chloride, biological reduction requires the addition of an electron donor, and most organic donors are expensive and pose potential health concerns through the formation of unknown biological reaction by-products. Additionally, organic donor residuals can contribute to biological regrowth in the distribution system.

Hydrogen is an ideal electron donor, as it is non-toxic, inexpensive, and sparsely soluble. It has not been widely used in water treatment, because it is difficult to deliver without sparging, which is wasteful and dangerous. Recently, a hydrogen-based hollow-fiber membrane biofilm reactor (MBfR) has shown promise as a safe and effective means to supply hydrogen to a biofilm process. In the MBfR, hydrogen is delivered directly to a biofilm growing on the membrane surface, resulting in safe and efficient hydrogen delivery.

In this research, initial feasibility investigations with the MBfR were performed to demonstrate its ability to reduce perchlorate. A major difference between the biological reduction of perchlorate and the more familiar biological denitrification processes is that perchlorate is typically present at concentrations orders of magnitude lower than nitrate. Consequently, it may be reduced as a secondary substrate, with another electron acceptor (nitrate or oxygen) serving as the primary substrate. For this to happen, perchlorate-reducing bacteria (PCRB) would likely have to be a subset of the denitrifying and oxygen-reducing microbial populations within the reactor.

During the initial investigations, the reduction of perchlorate to 4 μg/L from 60 to 1,000 μg/L was observed. No specialized inoculum for the MBfR was required, although a two-week adaptation period was required before removals up to 99 percent were observed. The lag presumably was the time required for enriching the perchlorate-reducing population. It was also discovered that the perchlorate-reducing ability of the MBfR was more sensitive to environmental conditions than its nitrate-reducing ability.

The presence of nitrate at levels as low as 14 μg-N/L significantly decreased perchlorate removals, but nitrate concentrations up to 200 μg-N/L had the same effect as 14 μg-N/L. pH control was important for perchlorate removal, as perchlorate reduction was more sensitive to the pH than was denitrification; pH 8 was optimal for perchlorate reduction. It was also discovered that perchlorate reduction could be lost without losing denitrification, suggesting that the microbial ecology of the reactor could be affecting the perchlorate reduction efficiency, and also that co-metabolic perchlorate reduction is not a significant perchlorate-removal mechanism.

Once it was established that the MBfR could be used to degrade perchlorate, kinetic studies were performed to better understand the fundamentals of perchlorate reduction under autotrophic hydrogen-oxidizing conditions. These tests were performed with a pure culture of *Dechloromonas* sp. PC1, an autotrophic hydrogen-oxidizing bacterium isolated from the MBfR. PC1 was characterized in terms of its phylogeny, physiology, and perchlorate-reduction kinetics. Chlorate, the first reduction intermediate from perchlorate, inhibits perchlorate competitively, a novel finding for a PCRB. When a high perchlorate concentration was reduced, enough chlorate accumulated to slow the perchlorate reduction rates. However, at low perchlorate concentrations, chlorate did not accumulate enough to have an effect on perchlorate reduction rates. Nitrate also has an inhibitory effect on perchlorate reduction, but it appears to be non-competitive and constant. PC1's K for perchlorate is 0.15 mg/L, around two orders of magnitude lower than found for other PCRB. Based on the kinetic parameters for perchlorate reduction, the estimated growth threshold for perchlorate (S_{min}), for autotrophic, hydrogen-oxidizing conditions, was around 5 µg/L. This means that perchlorate could not serve as a primary acceptor in a completely mixed, steady-state reactor with a 4-µg/L effluent perchlorate concentration, the California Department of Health Services Action Level. This finding underscores the need to have nitrate or oxygen present as a primary acceptor.

Further investigations into the microbial ecology of the mixed cultures in the MBfRs revealed that PCRB were present in a denitrifying system that had not been previously exposed to perchlorate. However, a dominant PCRB species increased from 14 percent to 21 percent of total bacteria when 100-µg/L perchlorate was added to the influent. Increasing perchlorate reduction led to further increases in the dominant PCBR and the perchlorate-removal capacity of MBfRs. Another important finding is that oxygen alone can serve as a primary acceptor for perchlorate reduction, and that the oxygen reduction appeared to be more favorable for perchlorate reduction than was nitrate reduction.

Scaling up the MBfRs by utilizing full-scale hollow-fiber membrane modules as the core of the pilot plant was the first step in understanding the long-term operational and maintenance issues associated with the operation of an MBfR. Also included in the pilot plant were an aeration basin to re-oxygenate the water after the MBfR and a dual-media filter to capture any sloughed biomass from the reactors. A total of four different MBfR reactor designs were evaluated as a part of this study. Through the testing of these reactors it was determined that the two most important factors affecting the long-term efficacy of the system were: 1) the mass-transport of perchlorate to the biofilm surface; and 2) the effective control and removal of excess biomass.

Changes to perchlorate mass-transport to the biofilm surface were realized by altering the water influent and effluent configurations, recycling water through the reactor, altering membrane fiber packing density, baffling, and removing excess biomass (to prevent channeling).

The effective control and removal excess biomass (i.e. cleaning the MBfR) was achieved through increased linear velocity through the reactor (recycle flow), increasing fiber mobility (only pot fiber on one end of the MBfR), regular air scour (daily), and periodic mild acid cleanings (monthly). These cleaning procedures were able to control the build-up of excess biomass that lead to clumping of the fibers, thereby reducing the effective biofilm area, exacerbating short-circuiting, reducing mass transport of perchlorate to the reductive organisms in the biofilm, and increasing the distance hydrogen must diffuse to reach the outer portion of the biofilm. Regular cleaning was also effective in minimizing sulfate reduction by impeding the proliferation of slow-growing sulfate-reducing bacteria. The cleanings were also shown to be an effective mechanism to reduce fiber breakage as a result of the calcification of the membrane fibers, or more specifically, the biofilm surrounding the membrane fibers.

Based on the knowledge acquired throughout the testing program, the anticipated cost of a full-scale MBfR based treatment system was compared to an existing perchlorate removal system. A wide range in the unit water production treatment costs associated with the scenarios evaluated was discovered. Perchlorate treatment (amortized capital, operation & maintenance) at the existing La Puente Valley County Water District 2500 gpm facility in southern California costs \$1.09/1000 gal (\$356/acre-ft). This system only removes perchlorate and discharges it into a dedicated brine line. Once perchlorate-laden brine discharge is no longer allowed, the additional costs associated with destruction of the perchlorate in the brine is anticipated to increase the costs of treatment to approximately \$1.32/1000 gal (\$430/acre-ft). While this represents a 21 percent increase in the cost of treatment, it is lower than the cost calculated for the initial MBfR designs used directly in the pilot study. However, projected modifications based on the results of the bench- and pilot-scale studies, along with good engineering, make the cost of an MBfR system that destroys perchlorate cost competitive at around $1.10/1000 gal ($360/acre-ft).

In addition to a technological and economic evaluation of the technology, multiple state agencies responsible for potential regulatory approval of the MBfR treatment process were contacted to obtain their comments about the MBfR process for the production of potable quality water. A variety of comments were received; however, the overall response summarized by the California Department of Health Services was that the, "membrane bio[film]reactor does appear to be a promising technology for perchlorate reduction. The Department looks forward to future communications from [the project team] regarding this technology."

APPENDIX A
ENGINEERING ANALYSIS CALCULATIONS AND ASSUMPTIONS

REACTOR DESIGN

The membrane design data is manufacturer and design specific and therefore critical to the input design section of the cost model. Included in this section are:

- Theoretical hydraulic retention time (HRT) ..... minimum of 15 minutes
- Bioreactor diameter .... 0.45 feet
- Bioreactor height... 4-feet
- Diameter of membrane fibers........................... 280 μm
- Length of membrane fibers .... 3.6 ft
- Number of fibers .... 8,000 (3.2% packing density)
  15,000 (6% packing density)
  50,000 (20% packing density)

For new systems, perchlorate flux rate will need to be provided by the membrane manufacturer or from pilot test data. The remaining membrane design data (listed above) should be available from the membrane manufacturer.

CAPITAL COSTS

The capital costs section includes equipment, annual O&M, labor costs, and construction related costs. As each of these are major sections, and therefore discussed individually.

Equipment

The equipment associated with a new MBfR plant includes:

- Initial membrane modules................................ $1500 each
- Replacement membranes .... $1500 each
- Re-circulation and effluent pump(s) .... $1000 per hp
- Aeration basin .... $1000 per hp (blowers)
- Multimedia filters.. $415,000 per MGD
- Facility structure .... $150 per square-foot

It is assumed for the purpose of the cost evaluation that the membranes will be replaced every 5 years and that all other equipment has a expected life of 20-years. This will vary with capacity, water quality, maintenance, and the amount of operation time (use rate).

Annual O&M Costs

Annual O&M costs are associated with daily consumable variables. These consumables include utilities as well as daily maintenance of equipment. For the cost analysis, the membrane life was assumed to be 5 years, with all other equipment life at 20 years. The O&M variables included are:

- Hydrogen costs.. $6 per Kg
- Cleaning costs.. $0.50 per pound of solution
- Energy costs... $0.12 per KW-hr
- Labor costs.. $30 per hour
- Maintenance costs... 1% of total capital costs

Labor Costs

The labor costs associated with operating an MBfR plant is estimated to be minimal once the initial plant startup is completed. This is due to system's ability to be automated, hence reducing daily operational requirements of the plant staff. The startup time will require the most operator attention and training to ensure proper installation, operating conditions, and that the biofilm is developing to an effective mass. The time is estimated at 2 weeks (full time) for 2 operators. Frequency of sampling and analysis requirements will depend upon design flow rate, and operating conditions (i.e. optimization period and plant upsets will have a higher sampling frequency).

Construction Related Costs

Construction related costs are those costs associated with construction, and for an experimental technology includes a risk factor. The analysis for the final report will evaluate a range of construction related costs (as a percentage of the capital cost), based on experience with water treatment plant construction. These ranges were based on experience with water treatment plant design and will vary with site specific information. Because the MBfR process is an innovative technology, similar full-scale projects are not currently available for reference. However, historical data is available for the auxiliary equipment, facility structure, and other more common construction related costs. The construction related costs are listed in Table A-1.

Table A.1
Range of construction related costs

Construction related cost	Average range*
Civil site work	1 to 10%
Instrumentation	3 to 15%
Electrical site work	7 to 12%
Piping	5 to 12 %
Construction contingency	10 to 35%
Engineering, legal and administrative	15 to 30%
Overall project risk	5 to 20%

* Average ranges based on experience with surface water treatment plant design.

Table A.2
Recommended contingency for corresponding level of estimate

Type of cost estimate	Level of accuracy	Recommended contingency
Order-of-magnitude	+50% to –30%	20% to 30%
Conceptual	+40% to –20%	20% to 15%
Preliminary design	+30% to –15%	15% to 10%
Definitive	+15% to –5%	10% to 5%

If the MBfR is located in a remote area, the range of construction costs may be higher than the data presented in Table A-1. For example, transportation and labor rates may be higher if labor and materials are not readily available in a remote location. However, we will assume that these are acceptable ranges for the analysis provided in the cost evaluations presented here.

Construction contingencies are often applied to the cost estimate to account for items not specifically included in a project scope but found to be necessary. The level of contingency selected should reflect the level of detail provided during pre-design. A low contingency budget reflects a high degree of confidence and a high contingency budget reflects a low level of confidence in the pre-design. A low degree of confidence may be due to limited data available for detailed costs, or the contingency may be higher if the treatment technology is experimental or new to the market.

There are four levels of cost analysis that depend on the amount of data available. Those four levels are order-of-magnitude, conceptual, preliminary design, and definitive. The recommended contingency levels for the varying types of cost estimates are listed in Table A-2.

MEMBRANE CALCULATIONS

Surface Area of Membrane Required. This number is based on the design flow rate, the perchlorate flux rate, and the difference between the influent and effluent perchlorate concentrations. Calculating the membrane surface area is critical as it is used to calculate the number of modules required for a given design flow rate and influent perchlorate concentration. The following values were used in the cost model:

- Influent perchlorate concentration .... 60 ppb
- Effluent perchlorate concentration .... 4 ppb
- Design flow rate .... 500, 1000, and 2500 gpm

The perchlorate flux was calculated using these input variables, see below.

$$\text{Flux} = \frac{(\text{Perchlorate}_{\text{influent}} - \text{Perchlorate}_{\text{effluent}})(\text{Flow rate})}{\text{Surface area}}$$

Total Number of Membrane Modules. This number is calculated by dividing the total membrane surface area by the membrane module surface area.

Packing Density per Module. The packing density is the number of fibers per module multiplied by the fiber cross sectional area, divided by the cross sectional area of the module. While the packing density does not effect the final treatment cost, it can effect the treatment efficiency (flux) which effects the final costs. This number has been included in the calculated design information to let the user be aware of how dense each module will be packed. An upper limit warning pops up if the density exceeds 10%. A recommendation will be provided based pilot testing observations.

Calculated HRT. The hydraulic retention time (HRT) is calculated in this section to verify the theoretical HRT entered in the input section.

Membrane Facility Structure. The building square footage is the footprint calculated from the sum of all equipment necessary to operate the MBfR plant that require enclosure. Once the facility footprint has been determined, the total square footage is multiplied by a unit cost of $150. This value is based on a one-story building and includes the costs of HVAC and lighting.

Membrane Cleaning Requirements. Pilot operation revealed that daily air scouring and monthly acid cleaning (citric acid) prevented excessive buildup of biomass within the MBfR modules. This section calculates the annual cleaning cost including compressed air requirements, chemical costs, and associated chemical cleaning equipment (i.e. clean-in-place unit, chemical storage, compressors, etc.). These variables are dependent on the number of bioreactors

Calculations to Determine Pumping Requirements

This section calculates the required horsepower per pump per system function (i.e. recirculation, backwashing, or final discharge). The cost of the pumps was calculated using $1000 per required horsepower. The horsepower requirements were calculated with the following equation:

$$\text{Horsepower} = \frac{r \cdot Q \cdot H}{550E}$$

where

r = Density of the fluid (water = 62.4)
Q = Flow rate (gpm)
H = Total head (feet)
E = Efficiency of pump (%)

A minimum of two pumps is recommended for all critical systems (i.e. recirculation, effluent, etc.). Based on design flow and recirculation rates, more pumps may be recommended to provide additional backup or system redundancy.

Pre- and Post-MBfR Treatment Requirements

The pre- and post-treatment equipment requirements for the MBfR plant include:

- Auto-backwashing strainer to remove suspended solids prior to entering the MBfR,
- Aeration requirements, and
- Post treatment filtering with multimedia filters

An auto-backwashing strainer is recommended as pretreatment to the MBfR for full-scale applications. A strainer would remove suspended solids prior to entering the MBfR and provide added protection to the system in the event of a raw water upset due to suspended solids. Aeration is recommended to provide dissolved oxygen (DO) to the treated water, as the system consumes DO during the treatment process. The multimedia filters are recommended to remove detached biomass that may discharge from the MBfR with the treated water

CONSTRUCTION AND OPERATION AND MAINTENANCE COSTS

This subsection presents the construction and O&M costs and the assumptions associated with these costs.

Construction

Assumptions Associated With Construction Costs

The following assumptions were made for the equipment and installation in the construction cost estimate (assumed included in cost):

- The bioreactors are low pressure systems complete with membranes, housings and all necessary support equipment,
- Each membrane units contains all necessary valves and connections,
- Installation includes integrity testing at each unit,
- Interconnecting pipe work complete with all necessary anti-vibration equipment,
- Motor control center (MCC) and distribution boards to supply all items of electrical plant - electrical system to include cabling, trays, trunking, accessories and supports between the membrane plant MCC, process control panel, motor drives and instrumentation,
- All necessary protective coatings and linings for plant,
- All ratings plates and labels,
- Feed and treated water quality instruments,
- All necessary trace heating and lagging, and
- Final effluent pumps.

Not included with the equipment and installation are:

- Feed water pumps,
- Feed water balance tank(s), and
- Filtrate water balance tank(s).

Present Worth Analysis

The present worth (present value) of the total capital costs to construct an MBfR plant, including the annual operation and maintenance needs, were evaluated by present worth analysis. The present worth of any value (such as the capital and O&M costs of the MBfR plant) is the equivalence of any future amount to the present amount. In this case it would be applied to the loan (or bond) estimate required to construct and operate the facility. The present worth is calculated by:

$$P = F(1+i)^{-n} = \frac{F}{(1+i)^{n}}$$

where

P = Present value
$(1+i)^{-n}$ = Single payment present worth factor
F = Future Amount
i = Interest rate
n = Number of years

Amortization

If the loan to construct the MBfR plant is to be paid back in equal payments over the life of the loan, the loan is amortized. The amortized cost for the MBfR plant was calculated through the division of the total capital cost over the number of years of the life of a loan:

$$\text{Amortized cost} = \frac{\text{Total cost (\$)}}{\text{Number of years}}$$

Water Production Cost

The water production cost is the cost per unit (i.e. per gallon, thousand gallons, acre-ft, etc.) to produce water. Understanding that preferences change between regions (including units) results are presented in both \$/1000 gallons and \$/acre-ft.

Membrane Cleaning

Given the influent water quality data for the existing MBfR pilot plant, daily air scouring and one mild citric acid cleaning per month has been necessary to control biofilm growth and prevent biomass accumulations that limit mass transfer. Included in the cost of membrane cleaning are number of cleanings per year, cost of cleaning solution (citric acid), and air scouring requirements (cfm).

APPENDIX B
REGULATORY PROCESS REVIEW

State of California—Health and Human Services Agency

Department of Health Services

California Department of Health Services

DIANA M. BONTÁ, R.N., Dr. P.H.
Director

GRAY DAVIS
Governor

JUN 2003
DHS-D.P.
CERTIFICATION UNIT
RECEIVED

June 16, 2003

Samer Adham, Ph.D.
MWH Global
555 East Walnut Street
Pasadena, CA 91101

Dear Dr. Adham: Samer

MEMBRANE BIOREACTORS FOR PERCHLORATE REMOVAL

Thank you for submitting your letter of March 17, 2003 (with the subject line: "Review of an Innovative Perchlorate Treatment Technology") detailing your work with membrane bioreactors and perchlorate reduction. As you are aware, the Department has accepted the use of biological treatment (fluidized bed reactors) for the reduction of perchlorate concentrations in sources used for the production of potable water supply. However, the Department has placed several stipulations on the use of biological treatment that might apply to your proposed membrane technology. The Department's Water Treatment Committee has reviewed your letter and has the following comments for your consideration.

1. Water production from the biological treatment process should minimize changes in production flow rates (e.g., a plant operated 24 hours a day, 7 days a week, 365 days a year to provide a minimum production of water (base loading or base flow)).
2. If variability in flow and composition for extended periods of time cannot be controlled and minimized, then product water should be stored to allow analysis before releasing the water to the distribution system.
3. Site-specific tests are required to determine the impact of seasonal and temporal variations in water quality (temperature, available micro and macro nutrients, etc.) on process performance. For example, it is anticipated the hydrogen requirement varies as a function of source water quality, so the impact(s) of variable nitrate concentrations (in time and magnitude) on finished water quality needs to be evaluated.

Drinking Water Technical Programs Branch, 2151 Berkeley Way, Room 458, Berkeley, CA 94704-1011
(510) 540-2158 FAX (510) 540-2152
DHS Internet Address: www.dhs.ca.gov Program Internet Address: www.dhs.ca.gov/ps/ddwem

4. Since the microorganisms are endemic to the groundwater, it is not necessary to identify and characterize the microbial population.

5. All chemicals used in the system must be NSF standard 60 certified by an ANSI accredited laboratory. Including the hydrogen gas.

6. It is recommended that all components used in the manufacture of the reactor vessel that come into direct contact with the source water be NSF standard 61 certified by an ANSI accredited laboratory.

7. It is also recommended that development continue on a reliable hydrogen control system that would allow feed-forward control of the hydrogen based on measured changes in composition and flow.

8. Treatment following biological perchlorate removal, at a minimum, may need to meet the disinfection requirements of the Surface Water Treatment Rule (Title 22 of the California Code of Regulations, Div. 4, Chapter 17) due to the possibility that opportunistic bacterial pathogens might survive and/or reproduce within the reactor vessel.

9. On-line monitoring systems for perchlorate and nitrate should be incorporated into process design for improving process control.

10. If VOCs, SOCs, or other organic compounds are present, it may be necessary to provide additional organics removal (e.g., advanced oxidation (UV/H_2O_2) and/or granular activated carbon) downstream of the MBR as an independent unit treatment process.

11. Designation as a best available treatment technology will require, by statute under the Health and Safety Code (Part 12 Chapter 4 Article 3 Section 116370), at a minimum, a full-scale application of the technology.

12. Your study is setup to demonstrate long-term steady-state operation, but what happens if there are short-term interruptions in the hydrogen feed or the perchlorate concentration? How will the treatment process be started after a power outage of a few hours? Non steady-state operation will be a critical issue. After the membranes are cleaned, how long will it take for the process to recover and start producing low perchlorate concentration product? This means looking at production stoppages and determining their impact (magnitude and duration) on product water quality (time dependent).

13. There are practical issues beyond the demonstration of the technology for perchlorate reduction. The residence time is stated as 7-30 minutes. How large must the system be (how many modules) to get any sort of production out of it?

14. Based on the kinetics, what would you say the minimum substrate concentration (perchlorate) would be, i.e., based on the biochemical reaction kinetics how much perchlorate would be left in the water?

15. The production of hydrogen sulfide is dismissed by stating that aeration and/or disinfection will readily oxidize it to sulfate. Theoretically, that is what happens. However, in actual practice, unless there is a large excess of oxidant, the oxidation is incomplete, producing intermediate thio compounds (e.g., tetrathionates) some of which cause objectionable taste and odor.

16. The comment is made that the process reduces perchlorate to chloride. Has this been demonstrated to be a 100% efficient reaction? If so, what was the evidence that provided the assurance that this reaction is going to completion?

17. What controls the maximum loading rate on the MBR, i.e., what limits or controls the efficacy of the MBR? How do perchlorate or other compounds impact the loading rate on the MBRs?

There may also be regulatory issues outside our Department's jurisdiction, such as worker health and safety with respect to the use and storage of hydrogen gas on-site.

The membrane bioreactor does appear to be a promising technology for perchlorate reduction. The Department looks forward to future communications from you regarding this technology. We hope these comments provide you with the feedback you are seeking and help direct future research so that the regulatory and permitting issues raised, can be addressed. Should you have any questions regarding the content of this letter, please free to contact me at (510) 849-5050.

Very truly yours,

Richard H. Sakaji, PhD, PE
Senior Sanitary Engineer

cc: WT Committee
chron

TEXAS COMMISSION ON ENVIRONMENTAL QUALITY

Protecting Texas by Reducing and Preventing Pollution

July 8, 2003

Mr. Samer Adham, Ph.D
Montgomery Watson Harza
301 North Lake Avenue
Pasadena, CA 91101

Subject: Review of Innovative Perchlorate Treatment Technology
Membrane Biofilm Reactor
AWWARF Project 2804

Dear Mr. Adham:

We wish to thank you for the opportunity to review and comment on the submitted perchlorate treatment technology report, dated March 17, 2003, using a membrane biofilm reactor. We wish to also apologize for the delay in our response. Our following responses are based on what the State of Texas shall require prior to installation of an innovative technology at one of our public water systems.

- While the Texas Commission on Environmental Quality (TCEQ) finds data from pilot studies conducted on waters outside the state informative, we have found that some of our raw water qualities have significantly affected the quantity if not the quality of the effluent from some innovative technologies. Therefore, we have historically required 90-day site-specific pilot studies. However, we will review a request not to conduct a site-specific pilot study based on the submittal of a comparison of the raw water qualities for the proposed location and the pilot study.

- The TCEQ has requirements that all additives to the water being treated and all material in contact with the water being treated must conform to American National Standards Institution/National Sanitation Foundation Standards (ANSI/NSF) 60 and 61, respectively, and be certified by an organization accredited by ANSI. An engineering representative of a public water system is required to provide such documentation with any proposal to pilot an innovative technology or at least documentation that they have submitted the required additives or materials for testing and shall have the required ANSI/NSF certification prior to installation of a full-scale unit.

P.O. Box 13087 • Austin, Texas 78711-3087 • 512/239-1000 • Internet address: www.tceq.state.tx.us

- Your letter states that the bacteria for the biofilm is grown from indigenous bacteria already present in the groundwater and will not be artificially inoculated or amended. We would request that the bacteria be identified. Also, we would like information on how long it takes to grow the required biofilm. What happens if the biofilm is killed by improper operation of the system? How long before the membrane biofilm reactor is able to produce an effluent that meets potable water requirements? The answers to these questions may require that systems have redundant units.

- The site-specific air scour frequency and duration a unit is out of service for removal of excess biomass in the membrane biofilm reactor should be monitored.

- We would want to know how the hydrogen gas is to be provided. Will it be generated on site? What type of quality control procedures are to be used to maintain the quality of the hydrogen gas being generated and the quantity being fed.

- We would require monitoring of the membrane biofilm reactor's loading rate (i.e., gpm/sf) for each site-specific pilot so that full-scale sizing can be verified. If alternative site data is accepted in lieu of a site-specific pilot, the loading rate would be limited to that piloted at the alternative site. At least one 30-day run should be conducted at the proposed full-scale loading rate.

- To meet the multiple barrier requirement, we are uncomfortable with the use of granular media being used as the filtration process for removal of the bacteria being grown on the hollow-fiber membrane. Also, the use of microfiltration membranes is questionable because of their absolute pore size. If hollow-fiber micro or ultra filtration membranes are proposed in lieu of the media filters, they would also have to be piloted to verify their operating parameters. As a second barrier, at least a 4-log removal of the bacteria and viruses would be required.

- The submittal mentions being able to detect fiber breaks from "condensation" caused by water flow into the lumen and displacement of the hydrogen. We were unclear as to where this condensation would occur and would need more data. If we did not find a comfort level with this process for detecting broken fibers, the TCEQ would require periodic direct integrity testing off line..

- The major area of concern for perchlorate in Texas at this time is noted for hard water. Based on the data reviewed, this may result in a permeant degradation of effluent quality and quantity over time. This consideration may result in the TCEQ requiring a pilot test period greater than 90 days or the installation of pretreatment units to lower the hardness level. Also, longer test periods may be needed to determine the best cleaning parameters on the membrane biofilm reactor.

- Disposal of the citric acid cleaning waste would have to be established and any required permits obtained. While neutralization and discharge to a wastewater collection system can be acceptable, the majority of our public water systems with elevated perchlorate levels do not have wastewater collection systems available.

- Also, some of the affected public water systems are going to have elevated levels of nitrates, arsenic, sulfate, chlorides and total dissolved solids. Based on the information provided in the submittal, monitoring for these constituents during piloting and periodic monitoring in full-scale units would be required.

- We did not notice sample results for the generated wastewater effluent's quality. These results would be needed for the TCEQ to evaluate any request for a discharge permit.

Thank you for the opportunity to review and comment on your research project. Again we apologize for our delay in providing our comments. Should you have any concerns regarding this letter, or if we can be of further assistance, please contact us at the letterhead's address.

Sincerely,

James 'Red' Weddell

James "Red" Weddell, E.I.T.
Technical Review & Assistance Team
Public Drinking Water Section (MC-155)
Water Supply Division

JSW:ad

cc: Mr. Joseph Strouse, P.E., TCEQ Utility Creation & Plan Review Team (MC-153)
Ms. Alicia Diehl, Ph.D., TCEQ Water Quality Monitoring Team (MC-155)
Mr. Geno Lehman, P.E., Montgomery Watson Harza

"Guterman, Damon (DEP)" <Damon.Guterman@state.ma.us>

07/09/2003 11:44 AM

To: <Geno.E.Lehman@us.mwhglobal.com>
cc:
Subject: RE: Innovative Perchlorate Treatment Technology

Geno,

I found the summary to be interesting and it does appear to be relevant to California action level (4 ppb) but the pilot data left 2 ppb in the effluent which given the (draft?) CA Public Health Goal of 2-6 ppb, the EPA draft assessment that suggested a safe level around 1 ppb and our own state's standard development process which is considering a 1 ppb level I wonder if MBfR can be optimized to produce an effluent level < 1 ppb?

The phrase "difficult to scour" areas as it relates to the need for more aggressive regulation of calcification leads me to ask if this complication can be predicted based on any other water quality parameter(s)?

Finally, it appears that the full treatment train may include MBfR, aeration, filtration (media or membrane) and disinfection. Given that there are competing pilots using GAC has a cost comparison been completed or contemplated as part of a larger pro/con analysis?

I realize these comments may come after the fact of the AWWARF report but hopefully they will prove useful in any future development of MBfR.

Sincerely,

Damon Guterman
Massachusetts Department of Environmental Protection
Drinking Water Program
1 Winter Street, 6th Floor
Boston, MA 02108
(617)574-6811
Damon.Guterman@state.ma.us

REFERENCES

Achenbach, L. A., U. Michaelidou, R. A. Bruce, J. Fryman, and J. D. Coates (2001). *Dechloromonas agitata* gen. nov., sp nov and *Dechlorosoma suillum* gen. nov., sp nov., two novel environmentally dominant (per)chlorate-reducing bacteria and their phylogenetic position. *International Journal of Systematic and Evolutionary Microbiology* 51: 527-533.

Alexander, M. (1981). Biodegradation of chemicals of environmental concern. *Science* 211(4478): 132-138.

Amann, R., W. Ludwig, and K.-H. Schleifer (1995). Phylogenic identification and in situ detection of individual microbial cells without cultivation. *Microbiological Reviews* 59: 143-169.

APHA, AWWA, and WEF (American Public Health Association, American Water Works Association, and Water Environment Federation) (1995). *Standard Methods for the Examination of Water and Wastewater.* 19th ed. New York: American Public Health Association.

APHA, AWWA, and WEF (American Public Health Association, American Water Works Association, and Water Environment Federation) (1998). *Standard Methods for the Examination of Water and Wastewater.* 20th ed. Washington, D.C.: American Public Health Association.

Aragno, M., and H. G. Schlegel (1992). The mesophilic hydrogen-oxidizing (knallgas) bacteria. In *The Prokaryotes: A Handbook on the Biology of Bacteria: Ecophysiology, Isolation, Identification, Applications.* Edited by K.-H. Scheifer. New York: Springer-Verlag. 344-384.

Ashelford, K. E., A. J. Weightman, and J. C. Fry (2002). PRIMROSE: A computer program for generating and estimating the phylogenetic range of 16S rRNA oligonucleotide probes and primers in conjunction with the RDP-II database. *Nucleic Acids Research* 30(15): 3481-3489.

Binder, B. J., and Y. C. Liu (1998). Growth rate regulation of rRNA content of a marine Synechococcus (Cyanobacterium) strain. *Applied and Environment Microbiology* 64(9): 3346-3351.

Binnerup, S. J., J. Bloem, B. M. Hansen, W. Wolters, M. Veninga, and M. Hansen (2001). Ribosomal RNA content in microcolony forming soil bacteria measured by quantitative 16S rRNA hybridization and image analysis. *Fems Microbiology Ecology* 37(3): 231-237.

Browner, C. M. (1999). Part II: Environmental Protection Agency. 40 CFT Parts 9, 141, and 142. Revisions to the unregulated contaminant monitoring regulation for public water systems; final rule. *Federal Register* 64(180): 50555-50620.

Bruce, R. A., L. A. Achenbach, and J. D. Coates (1999). Reduction of (per)chlorate by a novel organism isolated from paper mill waste. *Environmental Microbiology* 1(4): 319-329.

Bryan, E. H., and G. Rohlich (1954). Biological reduction of sodium chlorate as applied to measurement of sewage BOD. *Sewage and Industrial Wastes* 26(11): 1315-1362.

CaDHS (California Department of Health Services) (2001). *California's Experience With Perchlorate in Drinking Water.*

Chang, H. T., B. E. Rittmann, D. Amar, R. Heim, O. Ehlinger, and Y. Lesty (1991). Biofilm detachment mechanisms in a liquid-fluidized bed. *Biotechnology and Bioengineering* 38(5): 499-506.

Chaudhuri, S. K., S. M. O'Connor, R. L. Gustavson, L. A. Achenbach, and J. D. Coates (2002). Environmental factors that control microbial perchlorate reduction. *Applied and Environment Microbiology* 68(9): 4425-4430.

Coates, J. D., R. Chakraborty, J. G. Lack, S. M. O'Connor, K. A. Cole, K. S. Bender, and L. A. Achenbach (2001). Anaerobic benzene oxidation coupled to nitrate reduction in pure culture by two strains of *Dechloromonas*. *Nature* 411(6841): 1039-1043.

Coates, J. D., U. Michaelidou, R. A. Bruce, and S. M. O'Connor (1999). Ubiquity and diversity of dissimilatory (per)chlorate-reducing bacteria. *Applied and Environment Microbiology* 65(12): 5234-5241.

Cole, J. R., B. Chai, T. L. Marsh, R. J. Farris, Q. Wang, S. A. Kulam, S. Chandra, D. M. McGarrell, T. M. Schmidt, G. M. Garrity, and J. M. Tiedje (2003). The Ribosomal Database Project (RDP-II): Previewing a new autoaligner that allows regular updates and the new prokaryotic taxonomy. *Nucleic Acids Research* 31(1): 442-443.

Dahlen, E. P., and B. E. Rittmann (2000). Analysis of oxygenation reactions in a multi-substrate system—A new approach for estimating substrate-specific true yields. *Biotechnology and Bioengineering* 70(6): 685-692.

Dalton, H., and D. I. Stirling (1982). Co-Metabolism. *Philosophical Transactions of the Royal Society of London Series B-Biological Sciences* 297(1088): 481-496.

El Fantroussi, S., L. Verschuere, W. Verstraete, and E. M. Top (1999). Effect of phenylurea herbicides on soil microbial communities estimated by analysis of 16S rRNA gene fingerprints and community-level physiological profiles. *Applied and Environment Microbiology* 65(3): 982-988.

Engel, P. C. (1981). *Enzyme Kinetics, The Steady-State Approach.* Cambridge: Chapman and Hall.

Giblin, T., H. David, M. Deshusses, and W. Frankenberger (2000a). Removal of perchlorate in ground water with a flow-through bioreactor. *Journal of Environment Quality* 29: 578-583.

Giblin, T. and W. T. Frankenberger (2001). Perchlorate and nitrate reductase activity in the perchlorate-respiring bacterium perclace. *Microbiological Research* 156(4): 311-315.

Giblin, T., D. Herman and W. T. Frankenberger (2000b). Removal of perchorate from ground water by hydrogen-utilizing bacteria. *Journal of Environment Quality* 29: 1057-1062.

Giovannoni, S. J. (1991). *The Polymerase Chain Reaction. Nucleic Acid Techniques in Bacteria Systematics.* Edited by M. Goodfellow. Chichester, England: John Wiley & Sons. 177-203.

Hall, B. G. (2001). *Phylogenetic Trees Made Easy: A How-To Manual for Molecular Biologists.* Sunderland, MA: Sinauer Associates.

Herman, D., and W. Frankenberger (1999). Bacterial reduction of perchlorate and nitrate in water. *Journal of Environment Quality* 28: 1018-1024.

Herman, D. C., and W. T. Frankenberger Jr. (1998). Microbial-mediated reduction of perchlorate in groundwater. *Journal of Environment Quality* 27: 750-754.

Jackson, P. E., C. Weigert, C. A. Pohl, and C. Saini (2000). Determination of inorganic anions in environmental waters with a hydroxide-selective column. *Journal of Chromatography A*, 884, 175-184.

Joret, J., Y. Levy, and C. Volk (1991). Biodegradable dissolved organic carbon (BDOC) content of drinking water and potential regrowth of bacteria. *Water Science and Technology* 24:2:95-100.

Kengen, S. W., G. B. Rikken, W. R. Hagen, C. G. Van Ginkel, and A. J. M. Stams (1999). Purification and characterization of (per)chlorate reductase from the chlorate-respiring strain GR-1. *Journal of Bacteriology* 181: 6706-6711.

Kim, K., and B. E. Logan (2000a). Fixed-bed bioreactor treating perchlorate-contaminated waters. *Environmental Engineering Science* 17(5): 257-265.

Kim, K. J., and B. E. Logan (2000b). Microbial reduction of perchlorate by mixed versus pure culture. *Abstracts of Papers of the American Chemical Society* 220: U360-U360.
Koch, A. L. (1994). *Growth Measurement: Methods for General and Molecular Microbiology.* Edited by N. R. Krieg. Washington, D.C.: American Society for Microbiology. 248-277.
Koizumi, Y., J. J. Kelly, T. Nakagawa, H. Urakawa, S. El-Fantroussi, S. Al-Muzaini, M. Fukui, Y. Urushigawa, and D. A. Stahl (2002). Parallel characterization of anaerobic toluene- and ethylbenzene-degrading microbial consortia by PCR-denaturing gradient gel electrophoresis, RNA-DNA membrane hybridization, and DNA microarray technology. *Applied and Environment Microbiology* 68(7): 3215-3225.
Korenkov, V., V. Romanenko, S. Kuznetsov, and J. Voronov (1976). Process for purification of industrial waste waters from perchlorates and chlorates. U.S Patent.
Lee, K.-C., and B. E. Rittmann (2000). A novel hollow-fiber membrane biofilm reactor for autohydrogenotrophic denitrification of drinking water. *Water Science and Technology* 41(4-5): 219-226.
Lee, K.-C., and B. E. Rittmann (2001). Applying a novel autohydrogentrophic hollow-fiber membrane biofilm reactor for denitrification of drinking water. *Water Research* (Accepted).
Lee, K.-C., and B. E. Rittmann (2002). Applying a novel autohydrogenotrophic hollow-fiber membrane biofilm reactor for denitrification of drinking water. *Water Research* 36(8): 2040-2052.
Levenspiel, O. (1972). *Chemical Reactor Theory and Introduction.* New York: John Wiley & Sons.
Liu, W. T., T. L. Marsh, H. Cheng, and L. J. Forney (1997). Characterization of microbial diversity by determining terminal restriction fragment length polymorphisms of genes encoding 16S rRNA. *Applied and Environment Microbiology* 63(11): 4516-4522.
Logan, B. E. (2001). Assessing the outlook for perchlorate remediation. *Environmental Science and Technology* 35: 482A-487A.
Logan, B., and K. Kim (1998). Microbiological treatment of perchlorate-contaminated groundwater. *Proceedings of the Southwest Focused Groundwater Conference: Discussing the Issue of MTBE and Perchlorate in Groundwater.* 87-91.
Logan, B. E., H. S. Zhang, P. Mulvaney, M. G. Milner, I. M. Head, and R. F. Unz (2001). Kinetics of perchlorate- and chlorate-respiring bacteria. *Applied and Environment Microbiology* 67(6): 2499-2506.
Mackey, A. J., T. A. J. Haystead, and W. R. Pearson (2002). Getting more from less—Algorithms for rapid protein identification with multiple short peptide sequences. *Molecular & Cellular Proteomics* 1(2): 139-147.
Madigan, M. T., J. Martinko, and J. Parker (1997). *Brock Biology of Microorganisms.* 8th Edition.
Malmqvist, Å., and T. Welander (1992). Anaerobic removal of chlorate from bleach effluents. *Water Science Technology* 25(7): 237-242.
Malmqvist, Å., T. Welander, and L. Gunnarsson (1991). Anaerobic growth of microorganisms with chlorate as an electron acceptor. *Applied and Environment Microbiology* 57(8): 2229-2232.
Malmqvist, Å., T. Welander, E. Moore, A. Ternström, G. Molin, and I.-M. Stenström (1994). *Ideonella dechloratans* gen. nov., a new bacterium capable of growing anaerobically with chlorate as an electron acceptor. *Systematic and Applied Microbiology* 17: 58-64.
Manz, W., R. Amann, W. Ludwig, M. Wagner, and K.-H. Scheifer (1992). Phylogenetic oligodeoxynucleotide probes for the major subclasses of proteobacteria: Problems and solutions. *Systematic and Applied Microbiology* 15: 593-600.

Mayer, Kevin. (2003). Memorandum. Transmittal of Perchlorate Maps and Table. Updated April 2003. USEPA. May 14, 2003.

Miller, J. P., and B. E. Logan (2000). Sustained perchlorate degradation in an autotrophic, gas-phase, packed-bed bioreactor. *Environmental Science and Technology* 34(14): 3018-3022.

Miyazaki, S., H. Sugawara, T. Gojobori, and Y. Tateno (2003). DNA *Data Bank* of Japan (DDBJ) in XML. *Nucleic Acids Research* 31(1): 13-16.

Moeseneder, M. M., J. M. Arrieta, G. Muyzer, C. Winter, and G. J. Herndl (1999). Optimization of terminal-restriction fragment length polymorphism analysis for complex marine bacterioplankton communities and comparison with denaturing gradient gel electrophoresis. *Applied and Environment Microbiology* 65(8): 3518-3525.

Muyzer, G. (1999). DGGE/TGGE a method for identifying genes from natural ecosystems. *Current Opinion in Microbiology* 2(3): 317-322.

Muyzer, G., E. C. Dewaal, and A. G. Uitterlinden (1993). Profiling of complex microbial-populations by denaturing gradient gel-electrophoresis analysis of polymerase chain reaction-amplified genes-coding for 16s ribosomal-RNA. *Applied and Environment Microbiology* 59(3): 695-700.

Myers, R. M., S. G. Fischer, L. S. Lerman, and T. Maniatis (1985). Nearly all single base substitutions in DNA fragments joined to a Gc-clamp can be detected by denaturing gradient gel-electrophoresis. *Nucleic Acids Research* 13(9): 3131-3145.

Nerenberg, R., B. E. Rittmann, and I. Najm (2002). Perchlorate reduction in a hydrogen-based membrane-biofilm reactor. *Journal AWWA* 94(11): 103-114.

O'Connor, S. M., and J. D. Coates (2002). A universal immunoprobe for (per)chlorate-reducing bacteria. *Applied and Environment Microbiology* 68: 3108-3133.

Page, R. D. M. (1996). TreeView: An application to display phylogenetic trees on personal computers. *Computer Applications in the Biosciences* 12(4): 357-358.

Perciasepe, R. (1998). Part III: Environmental Protection Agency. Announcement of the drinking water contaminant candidate list; notice. *Federal Register* 63(40): 10273-10287.

Pernthaler (1997). *Methods in Microbiology: Marine Microbiology.* London: Academic Press.

Rikken, G. B., A. G. M. Kroon, and C. G. van Ginkel (1996). Transformation of (per)chlorate into chloride by a newly isolated bacterium: Reduction and dismutation. *Applied Microbiology and Biotechnology* 45: 420-426.

Rittmann, B. (1992). Microbiological detoxification of hazardous organic contaminants: The crucial role of substrate interactions. *Water Science and Technology* 25(11): 403-410.

Rittmann, B. E., and P. McCarty (2001). *Environmental Biotechnology: Principles and Applications*. New York: McGraw-Hill.

Rittmann, B. E., and P. B. Sáez (1993). Modeling biological processes involved in degradation of hazardous organic substrates. *Biotreatment of Industrial and Hazardous Wastes* 113-136.

Russell, J. B., and G. M. Cook (1995). Energetics of bacterial-growth—Balance of anabolic and catabolic reactions. *Microbiological Reviews* 59(1): 48-62.

Sáez, P. B., and B. E. Rittmann (1992). Model-parameter estimation using least squares. *Water Research* 26(6): 789-796.

Schmidt, J. E., and B. K. Ahring (1993). Effects of hydrogen and formate on the degradation of propionate and butyrate in thermophilic granules from an upflow anaerobic sludge blanket reactor. *Applied and Environment Microbiology* 59: 2546-2551.

Shields, M. S., S. O. Montgomery, P. J. Chapman, S. M. Cuskey, and P. H. Pritchard (1989). Novel pathway of toluene catabolism in the trichloroethylene-degrading bacterium-G4. *Applied and Environment Microbiology* 55(6): 1624-1629.

Shinoda, Y., N. Kato, and N. Morita (2000). Phylogenetic analysis of a bacterium by sequencing its 16S ribosomal RNA gene. *SHIMADZU Review* (in Japanese) 57: 121-132.

Spring, S., R. Amann, W. Ludwig, K. H. Schleifer, and N. Petersen (1992). Phylogenetic diversity and identification of nonculturable magnetotactic bacteria. *Systematic and Applied Microbiology* 15(1): 116-122.

Stahl, D. A., and R. Amann (1991). *Nucleic Acid Techniques in Bacterial Systematics. Development and Application of Nucleic Acid Probes.* Edited by M. Goodfellow. New York: John Wiley & Sons. 205-248.

Suzuki, M. T., and S. J. Giovannoni (1996). Bias caused by template annealing in the amplification of mixtures of 16S rRNA genes by PCR. *Applied and Environment Microbiology* 62(2): 625-630.

Teske, A., C. Wawer, G. Muyzer, and N. B. Ramsing (1996). Distribution of sulfate-reducing bacteria in a stratified fjord (Mariager fjord, Denmark) as evaluated by most-probable-number counts and denaturing gradient gel electrophoresis of PCR-amplified ribosomal DNA fragments. *Applied and Environment Microbiology* 62(4): 1405-1415.

Thompson, J. D., T. J. Gibson, F. Plewniak, F. Jeanmougin, and D. G. Higgins (1997). The CLUSTAL_X windows interface: Flexible strategies for multiple sequence alignment aided by quality analysis tools. *Nucleic Acids Research* 25(24): 4876-4882.

Tiedje, M. (1988). *Ecology of Denitrification and Dissimilatory Nitrate Reduction to Ammonium. Biology of Aanaerobic Microorganisms.* New York: John Wiley & Sons. 179-244.

Urbansky, E. T. (1998). Perchlorate chemistry: implications of analysis and remediation. *Bioremediation Journal* 2(2): 81-95.

USEPA (U.S. Environmental Protection Agency) (1993). Method 300.1 Determination of Inorganic Anions in Drinking Water by Ion Chromatography.

USEPA (U.S. Environmental Protection Agency) (1999). Method 314.0 Determination of Perchlorate in Drinking Water Using Ion Chromatography. Revision 1.0.

USEPA (U.S. Environmental Protection Agency) (2003). Ground Water & Drinking Water—Perchlorate [Online]. 23 January 2003. Available: http://www.epa.gov/safewater/ccl/perchlorate/perchlorate.html. [cited 11 March 2003]

van Ginkel, C. G., G. B. Rikken, A. G. M. Kroon, and S. W. M. Kengen (1996). Purification and characterization of chlorite dismutase: A novel oxygen-generating enzyme. *Archives of Microbiology* 166: 321-326.

Wallace, W., S. Beshear, D. Williams, S. Hospadar, and M. Owens (1998). Perchlorate reduction in a mixed culture in an up-flow anaerobic fixed bed reactor. *Journal of Industrial Microbiology and Biotechnology* 20: 126-131.

Wallace, W., T. Ward, A. Breen, and H. Attaway (1996). Identification of an anaerobic bacterium which reduces perchlorate and chlorate as Wolinella succinogenes. *Journal of Industrial Microbiology* 16: 68-72.

Ward, D. M., R. Weller, and M. M. Bateson (1990). 16s Ribosomal-RNA sequences reveal uncultured inhabitants of a well-studied thermal community. *Fems Microbiology Reviews* 75(2-3): 105-115.

Wilderer, P. A., H. J. Bungartz, H. Lemmer, M. Wagner, J. Keller, and S. Wuertz (2002). Modern scientific methods and their potential in wastewater science and technology. *Water Research* 36(2): 370-393.

Zhang, H. S., M. A. Bruns, and B. E. Logan (2002). Perchlorate reduction by a novel, chemoautotrophic hydrogen-oxidizing bacterium. *Environmental Microbiology* 4(10): 570-576.

Zumft, W. G. (1992). The denitrifying prokaryotes. In *The Prokaryotes: A Handbook on the Biology of Bacteria*. Edited by A. Balows, H. G. Trüper, M. Dworkin, W. H. Harder, and K. Schleifer. New York: Springer-Verlag. 554-582.

Zumft, W. G. (1997). Cell biology and molecular basis for denitrification. *Microbiology and Molecular Biology Review* 61(4): 533-616.

ABBREVIATIONS

A	electron acceptor substrates
ADP	adenosine diphosphate
AF	acre-foot
atm	atmosphere
ATP	adenosine triphosphate
BDOC	biodegradable organic carbon
BOD	biochemical oxygen demand
bp	base pairs
°C	degrees Celsius
CaDHS	California Department of Health Services
CCL	Contaminant Candidate List
CCPP	calcium carbonate precipitation potential
cfm	cubic feet per minute
Cl^-	chloride
ClO_2^-	chlorite
ClO_3^-	chlorate
ClO_4^-	perchlorate
cm	centimeter
CSTR	continuous flow stirred tank reactor
D	donor substrate
DGGE	denaturing gradient gel electrophoresis
DNB	denitrifying bacteria
DO	dissolved (aqueous) oxygen
DOC	dissolved organic carbon
DW	dry weight
FISH	fluorescent *in-situ* hybridization
g	gas
gpm	gallons per minute
H_2	hydrogen
HRT	hydraulic residence time
IC	ion chromatograph
K	half-maximum rate concentration

L	liter
LPVCWD	La Puente Valley County Water District
M	molar
MBfR	membrane biofilm reactor
MCC	motor control center
MDL	method detection limit
MDTC	membrane-covered dynamic thermal conductivity
mg/L	milligrams per liter
min	minute
mL	milliliter
mM	millimolar
MPN	most probable number
MRL	method report limit
μg/L	micrograms per liter
μL	microliter
ND	non detect
NH_4ClO_4	ammonium perchlorate
nm	nanometer
NO_2^-	nitrite
NO_3^-	nitrate
NWU	Northwestern University
O_2	oxygen
OD	optical density
OEHHA	Office of Environmental Health Hazard Assessment
PAC	Project Advisory Committee
PCR	polymerase chain reaction
PCRB	perchlorate reducing bacteria
PHG	public health goal
PNDM	perchlorate nitrate destruction model
Q	flow rate
QA	quality assurance
QAPP	Quality Assurance Project Plan
QC	quality control
Q_{max}	substrate utilization rate
R	recycle ratio
RfD	reference dose
RPD	relative percent deviation
RTD	residence time distribution

SEM	scanning electron microscopy
T	time
t_m	mean residence time
TOC	total organic carbon
t_R	theoretical hydraulic residence time
T-RFLP	terminal restriction fragment length polymorphism
TTHMFP	total trihalomethane formation potential
UCMR	Unregulated Contaminants Monitoring Rule
USEPA	United States Environmental Protection Agency
UV	ultraviolet
X	biomass concentration
Y	true yield

IWA Publishing's authorised EU representative for General Product Safety Regulations is Diane D'Arras, 15 rue Duret, 75116 Paris, France, e-mail: safety@iwap.co.uk.

Printed and bound by CPI Group (UK) Ltd, Croydon, CR0 4YY
17/09/2026
02218793-0019